A Fractured Profession

A Fractured Profession

Commercialism and Conflict in Academic Science

DAVID R. JOHNSON

Johns Hopkins University Press
Baltimore

Published 2017
Printed in the United States of America on acid-free paper
2 4 6 8 9 7 5 3 1

Johns Hopkins University Press
2715 North Charles Street
Baltimore, Maryland 21218-4363
www.press.jhu.edu

Library of Congress Cataloging-in-Publication Data

Names: Johnson, David R., 1977–, author.
Title: A fractured profession: Commercialism and conflict in academic science / David R. Johnson.
Description: Baltimore, Maryland : Johns Hopkins University Press, 2017. | Includes bibliographical references and index.
Identifiers: LCCN 2017003428 | ISBN 9781421423531 (hardcover) | ISBN 9781421423548 (electronic) | ISBN 1421423537 (hardcover) | ISBN 1421423545 (electronic)
Subjects: LCSH: Science—Study and teaching (Higher)—Economic aspects—United States. | Science—Research—Economic aspects—United States. | Universities and colleges—Research—Economic aspects—United States. | Universities and colleges—United States—Faculty. | College teachers—Professional relationships—United States. | BISAC: EDUCATION / Higher. | SCIENCE / General. | MEDICAL / Public Health.
Classification: LCC Q183.3.A1 J65 2017 | DDC 507.1/173—dc23
LC record available at https://lccn.loc.gov/2017003428

A catalog record for this book is available from the British Library.

Special discounts are available for bulk purchases of this book. For more information, please contact Special Sales at 410-516-6936 or specialsales@press.jhu.edu.

Johns Hopkins University Press uses environmentally friendly book materials, including recycled text paper that is composed of at least 30 percent post-consumer waste, whenever possible.

for Laura

CONTENTS

List of Tables and Figure *ix*
Acknowledgments *xi*

INTRODUCTION Professional Ideologies in Higher Education 1

1 Normative Tension in Commercial Contexts 21

2 The Reconstruction of Meaning and Status in Science 47

3 Embracing and Avoiding Commercial Trajectories 81

4 Identity Work in the Commercialized Academy 104

CONCLUSION Commercialism, Rationalization, and Fragmentation in Science 128

Appendix. Interview Protocol *147*
Notes *151*
References *163*
Index *173*

TABLES AND FIGURE

TABLES

1. Technology Transfer Activities, 2000–2007, Study Sample and AAU 17
2. Number of Research Articles, by Commercial Orientation and Rank 18
3. Selected Dimensions of Commercialization among Commercialists 19
4. Espousal of Disinterestedness and Universalism 22
5. Typology of Constraints to Communalism 23
6. Moral Orders of Commercialist and Traditionalist Science, by Dimensions of Work 48
7. Research Group Size among Chemists 50
8. Identity Work among Commercialists and Traditionalists 105
9. The Ideology of Commercialism 130

FIGURE

1. Average Salaries of Scientists 59

ACKNOWLEDGMENTS

I have long been fascinated by the meanings and rewards of work. For all the years I have devoted to thinking about these themes, never did I anticipate how satisfying it would be to thank the individuals and organizations who made this particular work possible.

My greatest debt is to Joe Hermanowicz, whose counsel never fails. For all my fortunes as a scholar, few have been as grand as working with Joe. This book is my thanks to him for his unparalleled guidance, intellectual support, and many readings of the manuscript since the project started. While at the University of Georgia, I had the privilege of working with Sheila Slaughter in the Institute for Higher Education. Perhaps more than anyone else, Sheila is responsible for the mountain of research that exists on the commercialization of higher education. Given her pioneering and continued work in this area, I am fortunate that Sheila helped me navigate toward the understudied and overlooked problems that are central to this book. I credit Jim Coverdill with encouraging me to study noncommercial scientists alongside commercialists and thank him and William Finlay for their close reading and insightful comments on the earliest versions of this work.

Much of this book was written during my postdoctoral fellowship at Rice University, where I worked closely with Elaine Howard Ecklund. Alongside the invaluable feedback she offered on this book, Elaine made considerable investments in my scholarly life and gave me the rare opportunity to expand my research agenda on a global scale. I will always cherish the fun that emerges through our collaborations. During this period, I was also fortunate to receive valuable input on this work from Erin Cech, Jared Peifer, and Brandon Vaidyanathan.

One of the greatest thrills of producing this book was sitting down with sixty-one scientists who are transforming how humanity understands and experiences the world around us. To these individuals and ten others who participated in the pilot study preceding this work, I extend my sincere gratitude for sharing with me in detailed terms their views and experiences in academe.

The broader study from which this book emerged was supported by a generous grant from the National Science Foundation. I was also fortunate to receive funding from the University of Georgia Graduate School and Rice University's School of Social Sciences. A National Institutes of Health grant (Sheila Slaughter, PI, with Maryann Feldman and Scott L. Thomas, co-PIs) generated the patent data that I relied upon to identify the population of academic scientists in the United States who patent their work. Other resources of value came from the Department of Sociology and the Institute of Higher Education at the University of Georgia, Rice University's Sociology Department, and the College of Education at the University of Nevada, Reno. To these organizations I extend my sincere gratitude.

I am very thankful to Greg Britton, executive editor at Johns Hopkins University Press, for the transparent and consistent support he provided throughout the review and publication of this book. I also appreciate the judicious copyediting provided by Barbara Lamb. Carolyn Bond did wonders copyediting this manuscript before I submitted it to Hopkins. I am also deeply appreciative of the insightful suggestions that I received during the review process of this book.

Although it may not be obvious to them, there are a number of individuals who shaped or supported this work in indirect ways. Mary Barnes and Anne Smith made early but critical investments in my writing. Jim Dodd, Randy Folsom, and Ernest Hebson taught me about social control, discipline, and coordination within organizations while teaching me music. Doug Mackaman, a fantastic historian of bourgeois culture and medicine in nineteenth-century France, exposed me to the intrinsic value (and unforeseen applications) of basic knowledge and spurred my initial turning point toward the professoriate. Vern Baxter took my interests in work and organizations and laid the earliest sociological scaffolding upon which all other doctoral training would rest. MaryAnne Lewis, Chris Moss, and Alec Watts each provided support for fieldwork at integral points in the study. Throughout my career, Scott L. Thomas has offered professional and scholarly advice, for which I am very grateful. Two of the most amazing individuals in my life (in vastly different ways) are Nick Conner and Bobby Siggins, whose friendship has always pushed me to be better in all endeavors.

My parents, Chuck and Linda Johnson, cultivated in me a deep appreciation for the good things that come from hard work and commitment. Both served the public good with distinction: my father as a pilot in the US Air Force; my mother as a public school teacher. The evidence of their success rests in the fantastic careers and wonderful families of my siblings, Charlie, Melinda, and Christina, whose spouses and children are a sheer delight. Our family weekends at Lake LBJ offered critical respite from the work of this book.

My greatest joy in life is my wife, Laura, whose love directs all that I do. Laura and I met not long before this project began, and her creativity, comedy, and quests for adventure have nourished my soul since the day we met. I could not have completed this book without her. Not long before I finished this work, our daughter, Arden, arrived in our lives. Thinking toward her future only serves as a reminder of the importance of science and higher education to society. Somewhere today an academic scientist is making a discovery, the technological importance of which will not become apparent until well into my daughter's adult life. I look forward to that moment with great anticipation and hope.

Professional Ideologies in Higher Education

Brian Hare, an evolutionary anthropologist at Duke University, started a company called Dognition, which, for $60, will analyze whether one's dog is a dunce or an intellectual prodigy. David Edwards, a Harvard biomedical engineer, created LeWhif, a company that sells inhalable chocolate, vitamins, and coffee. University of Georgia Professor Renee Kaswan invented Restasis, a drug to treat chronic dry eyes, which generated more than $75 million in royalty income and led to a prolonged lawsuit between Kaswan and the university.

Neither their potential superficiality nor the financial magnitude they involve makes these stories remarkable. Dognition allows consumers to assess the cognitive depth of their dogs, but the transactions generate data and funds for basic research. Although creating breathable chocolate may seem like unusual academic work, LeWhif is an offshoot of Edward's company Pulmatrix, which seeks to improve aerosol-based drug delivery. And while Restasis is a financial success—the most profitable invention to date from the University of Georgia —a blockbuster it is not. What *is* remarkable about these examples of research commercialization—the conversion of knowledge into products or services that can be sold—is the dramatic change they represent in the nature of work in academic science and the social organization of higher education.

Commercialization is by no means a new phenomenon, but it has radically accelerated since the 1980s.[1] Patents, the intellectual property rights assigned to inventions to facilitate technology transfer to the market, provide a reliable sign of this acceleration. Prior to 1981, academic scientists in the United States produced fewer than 250 patents annually. During the 1990s, the total number of patents owned by US universities increased from over 1,500 in 1992 to 3,000 by 1998. By 2003, universities had filed approximately 8,000 patent applications and completed 4,516 licensing deals that brought in an excess of $1.3 billion in royalties. The growth of spin-off companies was equally rapid: between 1980 and 2000, discoveries by university scientists generated 3,376 startup companies. Although industrial funding rarely represents more than 7 percent of funding for academic research, during the period 1972–2001 it grew more rapidly than any other source of research and development in higher education.[2] In 2012, industry provided $3.2 billion of funding for university research, with larger allocations going to highly ranked universities.[3] New forms of recognition have become easily observable in academe, as ranks, chairs, lectureships, awards, and university buildings frequently bear the names of corporations.[4] Higher education is in a period of unprecedented interface between the academy and industry.

University faculty—especially academic scientists—are now exposed to two main reward systems characterized by two different conceptions of the academic role and corresponding occupational norms, consequences for the advancement of knowledge, and rewards for achievement. Following sociologist Robert Merton, scholars once conceived of the scientific reward system as singular, referring to the *traditionalist*, or priority-recognition reward system, which mandates that scientists advance knowledge by sharing their discoveries with the scientific community through peer evaluation in exchange for recognition of priority in discovery. This honorary system of rewards now exists alongside a new *commercialist* reward system, which gives scientists a mandate to contribute to economic development through the dissemination of their discoveries in the market in exchange for profits.

These are not simply different approaches to scientific work. They are career paths tied to competing visions of the role of the university in society that raise questions with broad implications. What happens when—in a pattern following primary schools, prisons, and the military—the state retreats from the public sector and commercial interests rush in to fill that void? Why should taxpayers spend millions of dollars on research on the sexual behaviors of Japanese quail or the observation of distant solar systems, activities that have no obvious impact on society? What type of knowledge do we value as a society? Can we trust a professor funded by Monsanto to give unbiased information on consumption of poultry that is raised on genetically engineered feed? Should we trust, or even participate in, clinical trials when universities and scientists have a financial stake in the outcome of research? Isn't it a good thing that universities contribute to innovation and economic growth? Can this be done without undermining the quality of undergraduate education and graduate training?

In this book, I address the question, How does a commercially oriented reward system operate in the academic profession? The traditional role of universities as committed to a disinterested search for truth now exists alongside a new institutional goal of science that emphasizes the creation of technologies that have a concrete societal impact. The coexistence of these priorities is significant because many scientists and scholars believe that commercialization is antithetical to customary standards of scientific conduct and a threat to the pursuit of basic truths about the physical and social world in which we live. One logical way to answer this question would be to investigate how academic scientists perceive the role of commercial rewards in higher education. After all, this is where the action is: commercialization starts with academic scientists. Surprisingly, scholars have produced a wealth of research on the products of commercialization, such as patent activity, university startups, and university-industry linkages, but rarely does one find an in-depth and detailed examination of how commercialists themselves perceive this new reward system—a perspective I attempt to offer here.[5]

Another logical way to understand the implications of commercial rewards for higher education is to examine how academic scientists who *do not* commercialize their research view the commercial reward system. If commercialization is a threat to the production of basic knowledge or tantamount to the "corporatization of higher education," as some scholars allege, then it behooves us to ask how commercialization is *actually* viewed by "traditionalist" scientists who eschew it. Absent this perspective, we have only a loose grasp on the implications of commercial rewards for the academic profession, and we may be missing other consequences that scholars have yet to consider. But to read most scholarly research on the commercialization of higher education, one would conclude that the views and experiences of traditionalists in commercial environments are irrelevant to what needs to be known to understand the operation of commercial rewards in universities around the United States.[6]

This book attempts to provide a new understanding of the implications of commercialization for universities that is based on the narratives of sixty-one scientists at four elite universities in the United States. Focusing on commercialist scientists whose careers are indicative of this new reward system, and on their traditionalist peers in the same departments who eschew or reject commercialization, I argue that competing reward systems engender intraprofessional conflict. I develop the argument that the new institutional goal of science is *commercialism*. Commercialism is a professional ideology that asserts that scientists should create technologies that control societal uncertainties, wield power through major corporations, have a quantifiable impact that can be measured financially, and advance knowledge toward particular ends. Commercialism is fascinating in part because academics long viewed commercialized science as morally questionable and intellectually equivalent to making soap, yet now it is viewed by many as the hallmark of status and success. What is more, commercialism is predicated on values such as control, efficiency, calculability, and predictability, which many people associate with places like McDonalds, not academic scientific work, and yet science is by no means becoming rationalized into a routinized "McJob"—the scientists who embrace commercialism are among the most highly rewarded and autonomous scientists in the United States.

Commercialism has implications for traditionalists, who often find themselves embattled and threatened by universities' emphasis on commercialization. Traditionalists are less concerned about issues such as corruption, which outsider narratives suggest, than they are about unequal rewards, unequal conditions of work, and commitment to traditional goals of science and higher education, particularly the devaluing of basic research. This is suggestive of a new form of stratification in the academic profession tied to a conflict over the role of the university, how careers within it should be constructed, and how rewards should be allocated. These are centrally important findings with important implications for American research universities, because the resulting image is one of a fractured profession and thus of a tension—played out in the context of scientific careers—that is reshaping higher education and redefining what faculty are supposed to do. Before we arrive at this new understanding, however, we must first examine commercialization as a problem for the social organization of higher education.

Commercialization and the Social Organization of Higher Education

Commercialization of research is a chief manifestation of "academic capitalism," or market-like behaviors in universities and among faculty, such as patenting and licensing, spin-off companies, industrial funding, and student programs and services organized around economic returns. Scholarly attention to academic capitalism largely emerged with Slaughter and Leslie's *Academic Capitalism: Politics, Policies and the Entrepreneurial University*, which showed how politics and policies in the United States and elsewhere gave rise to intensification of market-like behaviors in higher education.[7] Subsequent attention shifted to organizational processes with Slaughter and Rhoades's *Academic Capitalism and the New Economy: Markets, State, and Higher Education*, which documented expanded managerial capacities within universities and mechanisms such as new circuits of knowledge and interstitial organizations that connect universities to the economy.

A voluminous body of knowledge on commercialization now exists, but consistent with academic capitalism's conceptual emphasis on organizational processes that link universities with the economy, most attention is devoted to the emergence of new organizational forms at the boundaries of universities designed to foster economic success. Roger Geiger and Creso Sá have detailed, for example, how federal and state governments coordinate with universities and industry to create centers for collaborative university-industry research organized around societally relevant research foci such as material science and nanotechnology.[8] Scholars have documented expansion in university managerial structures through the addition of technology transfer offices (TTOs) to seek interested companies to license faculty patents.[9] In my own work with Slaughter, Scott Thomas, and Sondra Barringer, we show how universities strategically enact ties to industry through their boards of trustees to foster patenting around shared technological interests.[10] Academic capitalism even poses new dilemmas for the organization of undergraduate and graduate education. A striking example is found in the work of David Kirp, who details the operation of a market mindset in University of Virginia administrators' decision to privatize the Darden Graduate School of Business in response to state budget cuts.[11]

This book provides one of the first in-depth looks at what these dynamics mean for faculty. Gary Rhoades recently pointed out that "far less common [in this area of research] is an understanding that academic capitalism, like capitalism more broadly, is about power embedded in the social relations of production."[12] In a similar vein, Slaughter asserts that researchers have largely ignored questions related to the diminishment of professorial power, the segmentation of faculty, and changing conditions of work in an era of academic capitalism.[13] Indeed, a central assumption of this book is that the careers of academic scientists are not only the chief basis for the production of knowledge—commercially relevant or otherwise—but also one of the key sites of power struggle and inequality in the academic capitalist order. There has been an absence of research that allows insight into how these dynamics are experienced by faculty.

Whereas Slaughter and Rhoades theorized changes in the *structure* of professional employment in higher education, my work theorizes changes in the *culture* of professional employment in the professoriate. The rise of a commercially oriented reward system entails the espousal of new rules for research, definitions of what is valued, career trajectories, and strategies for coping with changes in the allocation of resources. It also suggests the rise of segments of "haves" and "have-nots," either because commercialization may not be universally embraced by scientists or because only particular types of scientists are in a position to exploit this new reward system to their advantage. While some work points to the climates of contention this can create between faculty in the sciences and the humanities, this book takes us into how this dynamic unfolds in the commercial core of American universities: the social worlds of elite academic scientists.

Cultural Dimensions of Professional Organization

A profession's culture "consists of the ideologies that are relatively unique to it and the cultural forms that transform these abstract beliefs into a natural system."[14] Ideologies can be thought of as a profession's "oughts," which carry a moral significance crucial to definitions of appropriate behavior and the distribution of rewards. Ideology thus refers not only to belief systems but also to power. "To study ideology," says sociologist John B. Thompson, "is to study the ways in which meaning . . . serves to sustain relations of domination."[15] Historically, the traditional goal of producing socially certified knowledge has dominated American research universities to the extent that norms for research, the distribution of rewards for achievement, and how a successful career should look were relatively unambiguous. And even if this ideology of "traditionalism" was not universally agreed upon or observed in practice, the scientists and scholars who succeeded according to this system were those who accumulated the most rewards and influence.

To date, we know remarkably little about commercial culture in the academic profession. Scientists have not been entirely excluded from analysis, but overall most studies rely on methodologies that say little about what norms commercialists espouse, the values they assign to commercial practices, and why they embrace commercialism. Many scholars who do examine academic scientists are primarily interested in such problems as the formation of firms, why some regions are more innovative than others, or how particular markets emerge.[16] Consequently, we have no interpretation of the commercialization of academic research based on the perspectives of scientists themselves—those engaged in (or deliberately divorced from) this very commercialization; yet such an explanation holds the potential to transform how we conceive of commercialized academic research from understandings that focus on products of commercial practices to an understanding based on the experiences of the scientists.

Much remains to be understood about commercialists and traditionalists. While individuals never fall perfectly within ideal types, focusing on ideal types has value for the depth and detail it offers. I focus on commercialists and traditionalists to deeply explore the professional ideologies embraced by each group. Studying how traditionalists perceive and are shaped by the commercial reward system in depth is important because survey data indicate that about 90 percent of physical and life science professors fall into this category.[17] Studying commercialists who aggressively commercialize their research is important because their careers are indicative of the commercial reward system. The results of this comparative approach are powerful, as the study identifies more than twenty dimensions of comparison between the two groups. These dimensions, I show, demonstrate the rise of commercialism as a new professional ideology in higher education.

In the chapters that follow, I integrate research on the sociology of professions, science, and commercialization to motivate the development of a new perspective on the commercialization of science and the changing role of the university in society. This approach recognizes that commercialization is as much about professional ideologies of scientific work—embodied in espousal of norms, constructions of status, career trajectories, and professional identities—as it is about factors that other scholars emphasize, such as university policies, the presence of a technology transfer office, or network linkages between universities, industry, and government.[18] Much of what we have learned about commercialization is concerned with how to make universities more profitable rather than with the cultural dimensions of work, which allows us to study what it means, why different types of scientists think it matters, and what is at stake.

Norms

One of the chief though unresolved debates produced by the emergence of a commercially oriented reward system concerns the impact of commercialization on the normative structure of science. For Merton, the essence of scientific culture was found in norms of disinterestedness, organized skepticism, universalism, and communalism, which together constitute an ethos that guides and organizes conduct in science. Although inquiry into the operation and existence of norms once waned, contemporary scholars view them as a "touchstone" of debates over commercialization.[19] Consequently, one of the key questions I attempt to address is, How does a commercially oriented reward system influence the norms espoused by academic scientists?

Critics such as Derek Bok, Sheldon Krimsky, and Sheila Slaughter have noted that commercialization threatens the traditional normative structure of science.[20] The operation of the "Mertonian" norms, and the ways in which commercialized research is allegedly predicated on practices antithetical to these norms, may be seen in examining the kind of behavior each norm seeks to uphold.[21] Disinterestedness stipulates that the advancement of knowledge should proceed uninfluenced by scientists' stakes in the outcomes of their contributions, whereas commercialization directs the advance of knowledge toward commercial ends, motivated by the pursuit of profit. Organized skepticism refers to the social arrangements established to ensure that all scientific contributions undergo a fair and proper process of peer-based evaluation before becoming a part of certified knowledge. Under a commercial reward system, however, peer review is shifted from journals and departments to markets and utilizes criteria irrelevant to the advancement of basic knowledge. Communalism holds that scientific knowledge is a product of collective effort and must therefore be shared, not kept secret. By contrast, the new reward system places a premium on privatized intellectual property and could shift the nature of knowledge produced in universities from a public to a private good. Finally, universalism stipulates that scientific contributions, appointment, promotion, proposals for research funds, and honors are to be judged according to preestablished intellectual criteria, not according to particularistic social attributes irrelevant to scientific merit, such as personal characteristics, social status, or, in the case of present concerns, commercial relevance.

As I have documented in analysis of peer review in science, the so-called Mertonian norms are not norms indicative of most actual behavior; they are instead reflective of a rhetoric and ideology of science.[22] In the context of commercialization, norms that academic scientists *espouse* reflect how they think commercial rewards should or should not operate. Scientists, commercial and traditional, often disavow their actual behaviors (unseen by others) to uphold the status systems they construct. The norms that scientists espouse therefore reveal a great deal about ideologies of commercialism and traditionalism and the status systems associated with each.

Little more than anecdotal evidence was brought to bear on debates concerning commercial norms until Henry Etzkowitz's study of scientists at two universities in the early 1980s. The study, which was based on interviews with commercially oriented scientists, suggested that traditional scientific norms had been transformed by or replaced with an entrepreneurial ethos.[23] Although Etzkowitz's transformation hypothesis has gained favor among scholars, it is not without its problems. Scholars claim an entrepreneurial ethos has replaced the traditional scientific ethos "by dint of its prevalence," pointing to the increased level of commercial practices in academe, and characterize the "academic entrepreneur" as having a taken-for-granted status.[24] But this argument extrapolates from the behavior of a small subset of scientists: only 9 to 16 percent of academic scientists patent their research.[25]

In addition, scholars have arrived at this claim through an unfalsifiable analytic framework. If one studies *only* commercial scientists and finds that the norms they espouse systematically depart from traditional scientific norms, one could conclude that there are indeed new norms of science or that scientists are deviating from traditional norms.[26] But it is difficult to claim that the traditional ethos has changed or been replaced without verifying that noncommercial scientists in the same environments also no longer espouse traditional norms.

The claim lacks theoretic eloquence, too. First, scholars claim that an entrepreneurial ethos now exists without specifying what norms compose it. In the initial study from which the transformation-of-norms argument emerged, Etkowitz claims that new university linkages with industry "encourage normative change in science," but he did not specify any new norms.[27] The most specific articulation one can find in his work is in a subsequent article. Drawing upon 150 interviews at three points in time, Etkowitz states: "To put it in a nutshell, the new entrepreneurialism is the old one plus the profit motive. Seeking for funds has always been an important activity in the American research system, which demands a lot of entrepreneurial energy and phantasy. Therefore, as soon as traditional academic ambitions for the pursuit of the truth could be combined with profit seeking, the door was open for the new entrepreneurialism."[28] In the absence of identification of the new entrepreneurial norms, one can only assume they somehow correspond to behaviors seemingly antithetical to traditional scientific norms, such as secrecy and nonscientific motivations for research.[29]

The lack of knowledge on what values constitute commercial norms is consequential. The above studies suggest that commercial norms foster innovation, yet we cannot know *how* they foster innovation if we lack knowledge of *what* those norms are. Norms are also consequential to social organization because they operate as an arena in which contests over control of work occur. As sociologist of work Randy Hodson argues, "the struggle is not over who controls the workplace in the abstract. Struggles in the workplace are more often over specific norms and standards defining the nature of work."[30] Because research has yet to identify the values underlying a commercial norm or the extent to which such a norm may be contested by traditionalist scientists, we lack a complete understanding of normative change in science and the social organization of the academic profession.

Meaning and Status

Science, like other professions, can be thought of as a closed "social collective" with a distinctive meaning system. The meaning of scientific work is not simply a psychological concern for scientists pondering purpose in life. It influences the types of knowledge scientists produce, their investment (or lack thereof) in teaching undergraduates and training graduate students, how they view their commitment to the greater societal good, and critically, the status systems they construct. While much is known about the resources, interpretive schemes, and norms that have traditionally provided meaning concerning achievement in scientific careers, such attributes of the commercialist reward system are less well understood. A second research question central to this work therefore is, What are the implications of a commercial reward system for meaning and status in science?

To differentiate between commercialist and traditionalist science on this topic, we must consider how scientists conceive their role to understand the moral orders of science. As founding sociologist Emile Durkheim shows, moral order denotes the power of the collectivity over the individual, as collective beliefs, values, and orientations held by members of a group govern the actions and expectations of group members.[31] The fullest theoretical and empirical exposition of moral order in science is found in the work of sociologist Joseph Hermanowicz, who argues that "moral orders inform us about how careers are structured, interpreted, and understood."[32] Moral orders thus convey the aspects of scientific work that define commercialist and traditionalist science. Each order is a foundation for a different vision of science, which exhibits its own distinctive expectations, meanings, and practices that define group life.

Moral orders are important to the organization of work because they articulate the determinates of prestige and excellence and thereby operate as a key component of inequality. Moral orders in professions affect what types of task are considered the most difficult and therefore valued, which areas of knowledge are emphasized by the elite, and what it means to be exceptional. Traditionally, in science these dimensions of status are predicated on original contributions to research. Merton, for example, argues that the preeminence of original research contributions in determining status among scientists is a function of the centrality of the research role in relation to ancillary components of the scientific role set: "For plainly, if there were no scientific investigation, there would be no new knowledge to be transmitted through the teaching role, no need to allocate resources for investigation, no research organization to administer, and no new flow of knowledge for gatekeepers to regulate."[33] Because of the research role's functional centrality, "the working of the reward system in science testifies that [it] is the most highly valued. The heroes of science are acclaimed in their capacity as scientific investigators, seldom as teachers, administrators, referees, and editors."[34] Sociologist of professions Andrew Abbott's purity thesis expresses this argument in a more general theoretic form: "A profession is organized around the knowledge system it applies, and hence status within professions simply reflects the degree of involvement with this organizing knowledge. The more one's professional work employs that knowledge alone—the more it excludes extraneous factors—the more one enjoys high status."[35] Technologies and commercialization are ancillary to scientific knowledge—the basis of the profession—and thus the arguments of Merton and Abbott posit that fundamental research is the key source of status in science. Given that a commercial reward system is predicated on rewards for the invention of technologies, it would appear that the arguments of Merton and Abbott require elaboration. Without detailed evidence of the meanings that both traditionalists and commercialists assign to commercialization, and the bases of status they embrace, we understand poorly the patterns of social differentiation within the academic profession and the implications for the role of the university in society that stem from a commercially oriented reward system.

Career Paths

Changes in the career paths of scientists provide another dimension for assessing the implications of commercialization for the social organization of the academic profession. Here we are interested in the turning points at which scientists move into or avoid the newly recognizable status of "commercialist." We need to understand the social mechanisms that lead scientists to identify with, embrace, or reject commercial career paths. Does greed and self-interest fuel career turns toward commercialization? The absence of public funding? Are corporate interests luring scientists toward commercial career paths? We cannot really know the answers to these questions without examining how scientists account for their career paths. A third research question I examine, therefore, is, What are the social mechanisms by which scientists embrace and eschew commercialization?

The scholarly literature on determinants of commercial involvement contains relatively little information on scientists' subjective views about their work when they change from a traditionalist to commercialist career path, and no information about traditionalists' avoidance of the "path untaken." A major stream of research, for example, focuses on the importance of institutional properties, such as policies, or of characteristics of organizations, such as technology transfer offices and university administration.[36] These studies reflect Sheila Slaughter and her colleagues' theory of academic capitalism, which emphasizes the organization as a locus of social control and implies that responses to a commercial reward system are largely a function of resource dependency.[37]

A more recent stream of research suggests the need for understanding commercial turning points in the context of careers. A series of quantitative studies has examined characteristics of individual inventors and ways in which the decision to engage in commercial activity is conditioned by the local work environment. This research broadly suggests that scientists who patent are in mid-career, publish frequently, have trained and presently work at universities with large patent portfolios, and are more likely to be male than female.[38] Some work shows that scientists whose coauthors patent are more likely to patent themselves. Competing with such arguments are studies suggesting that scientists embrace commercial career paths in the pursuit of financial benefits, visibility, and status.[39]

Yet despite these studies, our understanding of commercial turning points remains limited. Many studies focus only on the transition to the first patent application or invention disclosure, but it is rarely true that such activities lead to commercial career paths. University policy requires scientists to disclose inventions, but it is technology transfer offices that decide whether patent applications should be filed, and most patents are not licensed.[40] We also know very little about how scientists account for commercial turning points in the context of their careers and how these accounts might vary among commercialists. Given that many scientists now engaged in commercialization received their PhD degrees before 1980, when the Bayh-Dole Act accelerated the onset of commercial culture in academe by making it easier for universities to claim intellectual property from faculty research, we need to know more about how mechanisms vary among scientists socialized to traditional scientific norms and those whose initial socialization to those norms included exposure to commercial practices. Comparison of career patterns and the ways in which scientists account for the timing of their commercial turning point is needed to understand how scientific careers have changed. Also needed is an investigation of the ways traditionalists explain why they avoid or do not pursue commercial opportunities. The motivations and processes by which faculty become commercially active, and by which traditionalism persists, thus require examination that situates those turning points in the context of careers.

Identity

The final research question I address asks, What are the implications of commercialization for professional identity? Many of the societal concerns about capitalism in universities—such as questions about corruption of research and the pursuit of self-interest—are tied to morality. Is commercialization the right thing to do? Or more precisely, the question for a scientist is, What is it good to be? Sociologist of work Everett Hughes famously asserted that "a man's work is one of the things by which he is judged, and certainly one of the more significant things by which he judges himself."[41] Hughes's point that work is a critical aspect of social identity is particularly true in science. Scientific work is deeply expressive of identity because it involves behaviors, ways of thinking, and the possession of esoteric skills shared with disproportionately few others in society.

A new reward system poses significant challenges for professional identity, or the collective identities scientists construct that define who they are, how they view themselves, and how they relate to and differ from other groups. Changes in the social organization of work, such as new divisions of labor, structures of work, and social relations within the profession, alter the possibilities for how scientists construct their identity through work. Structural change often benefits some workers while rendering others embattled or obsolete: "Many people," argues sociologist Robin Leidner, find that "economic change disrupts the attachment to place, job, organization, occupation, and career that had supported their identities. They are asked to embrace this new reality."[42]

Academic scientists, too, have had to embrace a new commercial reality, one that challenges the construction of identity. Cohorts of contemporary scientists who underwent doctoral training before the emergence of a commercial reward system developed professional identities in environments relatively free of referents to or opportunities for commercialization. In general, academic science before the 1980s exhibited a culture in which mixing the pursuit of knowledge with financial rewards was looked down upon.[43] This presents a challenge for the status passage from traditionalist to commercialist in two respects: First, it can be difficult to construct and enact a professional identity in a role transition that others may view as undesirable, for one's reputation and visibility within the scientific community may carry a stigma of moral taint. Second, the role transition can be problematic for the self: guilt or shame associated with the performance of tasks once perceived as morally questionable must be overcome both to sustain a positive self-identity and to persist in one's role.

While universities, government agencies, and industry promote the desirable benefits of commercialization, scientists also encounter ongoing reminders of the questionability of commercialization in a broad stream of accounts that are often highly publicized, such as official statements on the challenges that commercialization poses to research integrity; scholarship revealing instances of commercially related research that has endangered human well-being; publications in prestigious scientific journals that suggest commercialization constrains communication and produces research findings favorable to industrial sponsors;

congressional probes of commercially related research misconduct; and criticisms of close ties between universities and industry expressed by respected figures in the academic community.[44] These voices, not to mention similar accounts in major newspapers and trade periodicals such as the *Chronicle of Higher Education* merely highlight a chorus whose refrain implies that commercialization can be an undesirable practice. Even among commercialists who view their engagement with industry as desirable, there is a need to present a professional identity in a positive light.

The impact of commercialization on professional identity has implications for the social organization of higher education. Distinguishing between in-groups and out-groups is routine practice in society; we typically associate in-group distinctiveness with status and prestige.[45] Although academic science has never possessed a homogenous identity, even within disciplines or subfields, it is not difficult to imagine that the unequal distribution of new commercial rewards has caused the formation of competing professional identities, some of which scientists perceive as better than others. The fact that the only previous research on the relationship between commercialization and identity utilizes a sample skewed toward commercially oriented scientists hinders our ability to theorize the mechanisms by which scientists construct professional identity and therefore understand the implications of commercialization for the collective identity of the academic profession.[46]

Reward Systems Revisited

The classic perspective on science and professions depicts science as organized around the normative operation of its reward system. By definition, the pursuit of normatively sanctioned goals assumes consensus in the construction of institutional goals. According to sociologists such as Merton, a singular reward system fosters shared experiences, professional socialization, and standing in broader society. Collectively, these properties support and inform a mode of social organization predicated on shared norms, definitions of status, career paths, and professional identity. This Mertonian model, however, did not anticipate operation of two potentially competing reward systems.

What universities, like all societies and groups, do is contingent on conflict and contests. Commercialists and traditionalists can be viewed as actors who pursue status and power in conflict with others.[47] Indeed, Freidson notes that professions are characterized not by collegiality and trust but by dominance and power.[48] Examining academe, Pierre Bourdieu argues that universities should be viewed as arenas in which individuals and groups struggle to determine legitimate membership and legitimate hierarchy.[49] Contests for rewards entail different modes of power, yet we rarely get a glimpse at these processes in existing arguments about commercialization. The reasons for this are clear. Management scholars, who account for much of what we know about commercialization, have as their objective the identification of factors that promote innovation rather than issues of ideology and power in academe. Institutional theorists, on the other hand, emphasize how conformity to commercial rewards occurs, but they do so without an argument about competing ideologies, power, conflict, or the ruptures typical of organizational change.[50] Scholars of academic capitalism have theorized that commercialization could cause conflict between scientists (and faculty more generally) whose research is commercially relevant and those whose research is not, but they do not empirically explore the relations between these groups. And because scholars in general have yet to provide a detailed examination of perceptions of work in academic capitalist contexts, to date we have lacked an understanding of competing professional ideologies of work embraced by commercialists and traditionalists in these environments. This book begins to fill this gap through qualitative research designed to understand the dynamics and social relations among traditionalists and commercialists in the context of their careers.

The Research

During the spring of 2010, I conducted sixty-one in-depth interviews with scientists at four universities located in different regions of the United States.[51] The questions I asked scientists were divided into four interrelated sections: (1) conception of the scientific role; (2) motivations of commercialization; (3) norms of science; and (4) the operation of reward systems. The full interview guide is located in the appendix. The first three sections were designed to correspond to the specific research questions and related theoretic arguments I outlined above related to meaning, career paths, and espousal of norms, respectively. However, given the interrelated nature of the sections and the inclusion of open-ended questions, responses are relevant beyond the specific questions they were designed to address. Consequently, it was only through fieldwork and subsequent data analysis that the salience of professional identity to understanding commercialization became clear.[52]

I studied scientists who are situated at universities where the dynamics of commercialization are pervasive and a collective recognition of a culture of commercialization exists. It would be of no benefit to sample the views of scientists who are uninformed about the opportunity to commercialize one's work, for they would lack the opportunity to embrace, eschew, or meaningfully know about commercialization. My sample is thus not representative of all universities, but it *is* representative of the commercial culture at elite universities that lead the field of science and display commercial practices other universities seek to emulate.

To identify appropriate universities, I analyzed all of the patents assigned to academic scientists at the sixty-one US universities in the Association of American Universities (a consortium of elite universities) between 1960 and 2005.[53] Based on this and supplementary analyses, I adopted the following criteria for commercialists' inclusion in the study: five or more patents, founding of a company, or a licensing arrangement with an existing company. Based on the volume of commercialists who met these criteria, I selected two public and two private universities.[54] Each school is an elite research-intensive university at which exists a strong culture of commercialization. All four universities in the study are classified as Carnegie Doctoral/Research Universities–Extensive, a classification that reflects the broad range of baccalaureate and doctoral programs they offer in the humanities, social and physical sciences, and professional education. Total undergraduate enrollment at the universities ranged from 10,000 to 20,000 students at the private institutions and 30,000 to 40,000 students at the public institutions, while the total number of faculty at each school ranged from 1,200 to 1,700. Relative to the average AAU university, the universities in the study have higher numbers of graduate students and postdoctoral researchers and greater total research and development expenditures. Each of the schools receives substantially more money from industry than do its AAU peers.

In table 1, I present three indicators of technology transfer activities of the study sample and the average AAU university. To ensure the anonymity of the institutions in the study, I have averaged the total level of activity according to the public/private grouping of the schools. Across each indicator, the schools in the study substantially outperform the typical level of activity of AAU universities between 2000 and 2007. Relative to the average AAU university, both private schools in the study were assigned nearly three times the level of patents, about a thousand each, whereas the public schools received about one and one-half times more, approximately five hundred each. In terms of licensing agreements executed with new or existing companies, the universities generated more than five thousand licensing agreements between 2000 and 2007, roughly 20 percent of the total level of licensing activity among AAU universities during this period, with publics slightly outperforming privates. Relative to the average AAU university,

TABLE 1
Technology Transfer Activities, 2000–2007, Study Sample and AAU

	Publics avg. (schools 1 & 2)	Privates avg. (schools 3 & 4)	AAU average
Patents assigned	525	999	364
Licenses executed	1,461	936	446
Total license income (in millions)	$231	$360	$153

Source: Association of University Technology Managers, US Licensing Activity Surveys, 2000–2007.

Schools 3 and 4 received about twice as much licensing income, or about $360 million each, while Schools 1 and 2 received about one and one-half times the amount, or about $231 million each. Together, the four schools generated nearly 1.4 billion dollars in licensing income during the years 2000–2007. On each measure of commercial activity, each university tends to rank among the top fifteen universities in the AAU. These data confirm that the four universities in the study meet the sampling objective of selecting universities that exhibit a recognizable culture of commercialization.

Together, the four universities generated a sampling frame of 191 commercial scientists who met my criteria for inclusion in the study. For the sampling frame of traditionalists, scientists who had never applied for a patent, I randomly selected an equal number of traditionalists from each of the departments in which the commercialists were situated. When scheduling interviews, I sought a balance of scientists who received their PhDs before and after 1980, when the Bayh-Dole Act accelerated commercialization in academe, to ensure that I could study scientists who were socialized to traditional norms of science and transitioned to a commercial career trajectory. The overall rate of response to the study was 64 percent.[55]

The sample comprises thirty-two commercialists and twenty-nine traditionalists. Within each group, the sample is evenly divided across era of professional socialization and institutional type. Fifty-two of the scientists are male and nine are female. Among the females, two are commercialists and seven are traditionalists. Chemists make up the largest group of scientists in each category and account for approximately 41 percent of the sample, followed by the biologists, who compose about 31 percent of the sample. The remaining interviewees were drawn from chemical and biological engineering departments.

The achievements of commercialists and traditionalists in my sample are generally equivalent. By virtue of the scientists' appointments at four of the premier universities in the world, they share a level of excellence rare within the scientific community. The average participant had 150 publications as of 2010, but this

TABLE 2
Number of Research Articles, by Commercial Orientation and Rank

	Chemists		Biologists	
	Commercialist	Traditionalist	Commercialist	Traditionalist
Total publications				
By first appointment	8.9	9.7	6.4	5.5
At promotion to associate	22.9	26.9	15.4	12
At promotion to full prof.	42.3	38.7	32.4	27.1
n =	13	12	8	11

figure masks disciplinary and career stage differences. In table 2, I present the publication records of the chemists and biologists (the two largest groups in my sample) to control for field-specific differences in publication. At each career stage, the commercialists and traditionalists publish at roughly the same rate. The sample includes three Nobel laureates, sixteen members of the National Academy of Sciences, and eighteen fellows of the American Association for the Advancement of Science, with a slightly higher number of commercialists as recipients of these honors. All of the interviews were conducted on the condition of anonymity (at some points in this book I use fictitious names to illustrate a point).

The level of commercial activity among the thirty-two commercialists is important because it tells us about the experience from which these scientists' perspectives are derived. In table 3, I present selected commercial activities to illustrate this point. The number of patents per commercialist ranged from 0 (in the case of a chemist who started a company based on a copyright) to 250. All of the commercialists had entered licensing agreements. On average, 51.4 percent of commercialists' patents formed the basis of licensing agreements with new or existing companies. A total of 249 companies have licensed the work of the commercialists in the study. Seventy-eight of these companies (31 percent) are startups that were formed by the scientists in the study. Seven of the startups are now publicly traded corporations, and nineteen were acquired by or merged with other companies. All but six of the commercialists had started at least one company at the time of the interviews. Among these twenty-six commercialists, half had one startup company, nine had two to four, and four scientists had formed six or more companies.

Organization of the Book

In this volume I develop the argument that commercialization generates intraprofessional conflict in elite contexts of academic science. The book is organized

TABLE 3
Selected Dimensions of Commercialization among Commercialists

Activity	Average	Range
Patents invented	29.8	0–250
Number of patents licensed	10.6	1–80
Percentage of patents licensed	51.4	15–100
Companies licensed to	6.1	0–60
Startups founded	2.4	0–15

around the cultural dimensions of professional organization established in this introduction: norm espousal, meaning and status, career paths, and identity. Because beliefs about how work should be performed and rewarded are the central components of professional ideology, I begin with norm espousal and meaning and status. Norms have also been central to most debates about a profit motive in science. I therefore begin the book with chapter 1, "Normative Tension in Commercial Contexts," which is organized around the traditional norms of science to assess the extent to which commercialists and traditionalists espouse adherence to longstanding norms of science. The findings specify the circumstances under which commercial activities and commitments are in tension with the traditional "rules" for how science should be done. The central themes the chapter considers are whether scientists should be concerned with the utility of their research, whether scientists should be rewarded for commercial activities, and the contexts of secretive behaviors connected to commercialization.

Chapter 2, "The Reconstruction of Meaning and Status in Science," builds on chapter one by examining the status systems associated with the "rules" of science developed in the first chapter. To do so, I examine the influence of commercialization on meaning and status in scientific careers. I examine four dimensions of scientific work that constitute moral orders of commercialist and traditionalist science: the logic and organization of scientific work, the status scientists seek, the reputations they attempt to construct through visibility, and the activities they avoid. Through these comparisons one sees new definitions of prestige and excellence in the academic profession.

Having established the key components of ideologies of commercialism and traditionalism, I turn to the processes by which scientists were socialized to commercialization and the identity work they perform in academic capitalist contexts. In chapter 3, "Embracing and Avoiding Commercial Trajectories," I discuss the social mechanisms by which scientists embrace or forego commercial trajectories. All professions, and the diverse career paths within them, exhibit distinctive attractors and facilitators—characteristics that offer benefits and impel individuals in one direction or another. This chapter illustrates the attractors and facilitators that emerged in scientists' accounts of their career trajectories. First I examine why some scientists became commercialists by considering the influence of factors such as training, luck, money, and ethics and how certain factors are unique to junior and senior faculty cohorts. Next, I consider why traditionalists remain traditionalist in highly commercial contexts and even after being approached by corporate scientists. I consider influences such as opportunity structure, commitment to scientific norms, and goal incompatibility.

Even within a stable commercialist or traditionalist career path, the maintenance of one's professional identity is an ongoing process. For commercialists this is because they must present themselves to their peers as morally legitimate faculty and not money-motivated "corporatized" scientists. For traditionalists this is because they need to distance themselves from the commercial culture that surrounds them. In chapter 4, "Identity Work in the Commercialized Academy," I consider the implications of commercialization for professional identity. I begin by examining processes of legitimation and demonstrating four techniques that commercialists employ to pursue profits without appearing to be morally questionable. Then I examine processes of disidentification, demonstrating four techniques that traditionalists employ to affirm the purity of their identity.

In the conclusion, "Commercialism, Rationalization, and Fragmentation in Science," I draw on the main findings of the book to summarize and explain the key components of the ideology of commercialism and how it is distinct from traditionalism. I show how elements of substantive rationality that place a premium on societal impact through economic behavior are now deeply ingrained in commercialists' understanding of how science should be done. Commercialism is predicated on the control of uncertainty, the optimal use of power to exercise societal influence, the quantification of scientific impact through the market, and the progress of science toward targeted societal problems. I consider more broadly how commercialism results in intraprofessional conflict in science, its implications for fragmentation of the academic profession, and why this matters for the future of the American research university.

Normative Tension in Commercial Contexts

Groups have norms, or rules about how one should act. According to sociologist Erving Goffman, a social norm "is that kind of guide for action which is supported by social sanctions, negative ones providing penalties for infraction, positive ones providing rewards for exemplary compliance."[1] Norms reflect a group's ideals, and although ideals are rarely indicative of how most members of a group behave, they do reflect beliefs about how a system should operate. The norms that a scientist espouses therefore form the basis of a professional ideology of how she believes a scientist should act and of her vision of the role of research universities in society.

In this chapter, I treat the norms that scientists espouse as central to the professional ideologies of commercialism and traditionalism. Three of the four "Mertonian" norms discussed in the introduction—disinterestedness, universalism, and communalism—emerged in my interviews with scientists.[2] In contrast to studies that pose questions directly related to these norms, I did not set out to test the veracity of Merton's claims. Instead, I employed a set of open-ended questions that addressed scientists' perceptions of the appropriateness of commercial behavior.[3] Even so, my analysis of the interview data affirms (likely to the chagrin of those who see Merton's analysis of norms as outdated) that disinterestedness, universalism, and communalism remain significant to how contemporary academic scientists view their roles. Yet, traditionalism bears an increasingly tenuous hold on the work of many scientists at elite universities, a grip weakened by the powerful influence of commercialism in these environments.

Through my interviews with scientists, I found that traditionalists espouse the institutionalized norms of disinterestedness, universalism, and communalism, whereas commercialists attitudinally or behaviorally reject them. The commercialists' rejection and the traditionalists' espousal of disinterestedness—the prescription that research should be done for the sake of science—is observable in

TABLE 4
Espousal of Disinterestedness and Universalism

	Commercialists	Traditionalists
Disinterestedness		
Problem selection	Utility	Scientific criteria
Reference groups	Industrial	Academic
Universalism		
Perception of commercial rewards	Market premium	Caste system

the ways they make decisions about which problems to work on and the reference groups to whom they are committed (table 4). As I attempt to show, commercialism is predicated on formally rational elements of efficiency, calculability, and predictability in harnessing industrial resources and power to seek utilitarian solutions to particular societal problems. Traditionalists stress that scientists should conduct research that will be useful primarily in terms of its impact on knowledge and the scientific community. They reject the notions of utility and industrial collaboration that commercialists favor with regard to reference groups and the selection of research problems, arguing that most of the important technological developments that have had major societal impacts are products of disinterested pure research.

A similar pattern of rejection and espousal, respectively, characterizes the orientation of commercialists and traditionalists to the norm of universalism, which stipulates that scientific honors should be rewarded according to scientific, not social, criteria. Both commercialists and traditionalists believe that the royalties derived from commercialization are legitimate, but they diverge regarding the role of commercial success in how professional rewards are allocated. Traditionalists view commercial success as irrelevant to advancing knowledge and see university incentives for commercialization as contributing to a caste system within universities. Commercialists construct the market as the ultimate form of peer review and hence argue that marketplace success is a legitimate reason for receiving career rewards.

Commercialists and traditionalists do not diverge in their rejection of secrecy, implying adherence to communalism—the expectation that scientific knowledge is to be shared because the growth of knowledge is a collective effort. Even so, most of the scientists I interviewed acknowledged that commercialization does in fact undermine the norm of communalism. Most of the instances of secrecy reported were performed by commercialists, although commercialist interviewees admitted only minor acts of secrecy related to temporary publication delays.

TABLE 5
Typology of Constraints to Communalism

Context	Secretive Behaviors
Peer review	Poaching and sandbagging
Conference presentation	Posturing and positioning
Graduate training	Restriction, rerouting, and misappropriation
Publication	Delaying and withholding

My interviews with scientists of both groups revealed four contexts in which they observed secrecy and corresponding behaviors: peer review of research and proposals, conference presentations, graduate training, and publications (table 5).

Disinterestedness

The norm of disinterestedness governs motivations for research and thus provides institutional control over which problems are selected for research. The basic prescription of this norm is that research should be done for the "sake of science," not for the sake of rewards. In other words, research and discovery are ends in themselves, rather than means for acquiring professional recognition, influence, or financial success. The importance of this norm is that it emphasizes that only scientific criteria should determine what research one undertakes. Disinterestedness is thus critical to the reward system because scientists obtain rewards for having solved problems of collectively agreed-upon significance to the advance of knowledge. Let us consider the impact of commercialization on this norm, in terms of two dimensions of disinterestedness that emerged in my interviews with scientists: the criteria by which scientists select problems and the reference groups to whom they are committed.

Problem Selection: Societal Utility versus Scientific Criteria

Although Merton did not explicitly discuss the relationship between disinterestedness and criteria for selecting problems, doing research for the "sake of science" implies that only certain criteria are consistent with this institutional imperative. A faculty member who embraces disinterestedness would consider criteria such as consensus among colleagues about which problems are important or what projects could potentially resolve problems at the frontiers of knowledge in one's field. By contrast, selecting a problem because it has technological and financial implications may not necessarily advance basic science and cannot exclusively be viewed as pursued for the sake of science.

In the view of commercialists, the key criterion in selecting a problem is utility—that is to say, whether or not it is materially practical. Commercialists frame their commitment to commercially oriented research in terms of a moral obligation to society and the funders of scientific research. According to this view, the question a scientist should pose when selecting a problem is, What material impact could this research have on society? I interviewed "Rose," a professor of chemical engineering in her forties,[4] who started a company with one of her postdoctoral scientists. In a small office, where hung a picture of her shaking hands with a former US president, she explained:

> I think that when the funding is coming from a source such as the government that there should be an end to the query. So I think that when we talk about dark matter, that's a good place, right? And supercolliders. I think that there has to be an appreciation for why that science is important. The people who write those proposals need to be able to lay out the argument for that. As long as you can lay out a reasonable argument for a mode of inquiry that leads us to some greater understanding and it's not just because "Why?" I think that's reasonable.

Notice that this commercialist scientist emphasized not the intrinsic value of research and discovery but rather the idea that academic scientists need to justify the social merits of the results of research problems they choose. For her, asking "Why?" for its own sake—a question indicative of the unique role of universities and disinterested science—is insufficient justification for a research choice. Here we see the operation of a formally rational value of predictability in that commercialism entails an emphasis on pushing knowledge toward particular targets, such as renewable energy or human health and well-being. It does not mean that commercialists do not engage in curiosity-driven research. Rather, commercialism encourages connecting curiosity-driven research and problems that cause uncertainty in society.

For commercialists, traditional outcomes of disinterested science, such as contributions to the growth of knowledge and training of future scientists are, alone, no longer viewed as worthy products of science in that they do not directly lead to a materially quantifiable economic benefit. Consider, for example, a comment by a scientist who,[5] when asked whether scientists should be concerned with the utility of their research, stressed calculability, or the need to be able to quantify scientific impact:

> Agencies and companies that fund research want to measure return on investment, and it's very hard to measure it when the outcome is published papers, graduating PhD students, and things like that. It's much easier if an agency looks at the stuff they were funding here and said we were supporting the scientists that founded [an IT company] and they can point to a huge impact, an economic impact. That is a valid thing to do. At the end of the day, if all we did was support research in which there is never any connection with the commercial world, then that's swinging the pendulum too far the other way.

The core value underlying utility is calculability. In other words, the utility factor can actually be calculated, first in the form of a technological product or a company and then in the economic and societal impact of that product. As a result, for commercialists, utility is not simply a criterion in problem selection; it is the hallmark of excellence in scientific research. The commercialists I interviewed see utility as enriching science, and they view the market as a form of peer review whose results can be measured by the breadth of impact of one's work.

If the entrepreneurial norms inherent in commercialism have transformed, undermined, or displaced traditional scientific norms for academic scientists, as some scholars assert,[6] we would expect traditionalists, like commercialists, to consider utility a key criterion in the selection of research problems and the merits and goals of industrially sponsored research. In fact, the most common response when I asked traditionalists about utility as a basis for choosing a research problem was dismissal of that idea, combined with a comment about how the important transformative findings are arrived at through research done purely for scientific reasons. The notion that research and discovery are intrinsically valuable pursuits independent of a utilitarian end was almost invariably present in traditionalists' accounts of their work. As a professor of biology in his mid-forties and I were discussing whether or not scientists should think about practical aspects of their work when selecting research problems,[7] he told me:

> There are lots of things that I can point to that people did that seem very self-serving and not of any application [at first blush, but that eventually] became critical. One of my favorite examples is cell cycle control. Lee Hartwell decided it would be cool to understand how yeast cells divide. He basically said, "I bet I can get mutants that mess this up" . . . [And now] most of what we understand about cancer stems from his fundamental discovery of how a yeast cell divides . . . If he had to apply his research, he probably would have worked for Budweiser, right? There's an intrinsic value to letting people explore. I think anybody who forgets that is not learning from history.

Another traditionalist professor of biology,[8] a member of the National Academy of Sciences, when discussing the same topic, similarly appealed to the transformative impact of fundamental science. However, while the biologist in the previous example emphasized the possible positive societal impacts of basic research, this biologist pointed out that scientists should not be concerned with possible negative impacts of their work:

> Should Einstein have been worried about how his understanding of nuclear physics might have generated a nuclear weapon? No. What he did had beauty and value intrinsic to the discovery, and that's true of any basic science. I certainly don't feel that one should be concerned, in designing a course of research, about its application. Not at all. The university would fundamentally change if that were a consideration.

Both of these comments imply that the pursuit of scientific knowledge in universities should be uninfluenced by nonscientific criteria, whether positive or negative, a point tied to their vision of scientific growth. The notion that the university would "fundamentally change" if such factors influenced research choices reflects the degree to which traditionalists see professional self-regulation, free of external influence, as critical to the advancement of knowledge, a recurring point in traditionalists' accounts of science.

Reference Groups: Industrial versus Academic Influence

The second dimension of disinterestedness that emerged in the interviews was reference groups. The notion of reference groups, or the groups scientists seek to influence and receive rewards from, is tied to the norm of disinterestedness because whether a scientist is influenced by internal (academic) or external (industrial or corporate) reference groups determines the types of criterion he or she employs in problem selection.[9] There are two possible relationships between disinterestedness and reference groups. The operation of disinterestedness will be undermined if the selection of research problems is influenced by external reference groups, for such an arrangement intermingles scientific with nonscientific interests such as money, politics, or religion. Alternatively, we might assume that industry-funded research perfectly overlaps with the scientific criteria the academic community uses to designate which problems are worthiest of investigation, in which case disinterestedness will be sustained.

My interviews revealed that commercialists are highly committed to industrial reference groups beyond their universities. To illustrate this orientation and its influence on disinterestedness, we can consider how a professor of chemistry[10] framed a major industrial liaison program between his university and one of the largest pharmaceutical companies in the United States. When I asked him if he saw problems associated with scientists accepting research funding from corporations, he replied no and explained:

> I think the system at [this university] has to be modeled as a way in which the faculty members have the privilege or option of applying for these arrangements. They have a research idea and they ask [a pharmaceutical company], "Are you interested in investing some money in this research idea?" And then [the pharmaceutical company] and [the university] as a group take a bunch of ideas and say, "These are the ones that we think are valuable, we will make those investments." It isn't a company telling [the university or] faculty members what to do; it is a company saying, "We'll make judgments with our colleagues at [this university] from a company's perspective, from a pure intellectual or not so pure intellectual perspective as to what's valuable, and we'll select what we're going to do with these resources."

This scientist's comment dismisses the possibility that the interests that motivate industrial sponsorship of research and the criteria by which academic scientists select research problems naturally overlap. That is, he acknowledges that the selection process may be from a "not so pure intellectual perspective." Such an acknowledgment does not, of course, suggest that the motives of companies that fund academic research are sinister. That would miss the analytic point that merits our attention. Disinterestedness is presumably undermined because nonscientific criteria compete with scientific criteria in the determination of the problems on which faculty should work. Faculty I interviewed clarified that the industrial consortia or specific industrial sponsorship projects at their universities were not providing general research funds for the universities to distribute across all colleges or fields. Corporate financial contributions to universities are thus motivated by some form of interest, and yet it is true that companies have historically provided unrestricted funds to scientists to use as they see fit. The accounts of most scientists, however, suggest that, more often than not, industrially sponsored research is driven by the research concerns of industry, not of the academic scientific community.

Given that corporate sponsors have specific technological interests they wish to promote, the level of influence of nonscientific reference groups on academic research is ultimately shaped by how commercialists negotiate sponsored and collaborative arrangements with their corporate sponsors. The commercialists I interviewed often characterized these corporate collaborations as synergistic partnerships, to the extent that sponsorship facilitated their own research agenda, but some commercialists described scenarios rejecting a company's interests for the sake of scientific autonomy. I interviewed a professor of chemistry in his late forties whose vita indicates that, during his nineteen years as a professor,[11] he has received a total of ten years of funding from chemical, pharmaceutical, and medical companies. He explained:

> I have multiple companies come and say, "We'd like to fund a postdoc." And I say, "I want to be free to choose what research this postdoc does. You could give me the money, we could do some research for a year, give you a report of what we have done, and if you like it, you could try it again." But that's not what they want. They want to tell me what experiments they want done, and that is just totally abhorrent to me. But it's not abhorrent because it's ethically abhorrent; it's abhorrent because I personally want the freedom to do what I do. I don't want people telling me what to do. I want to be my own boss.

Here we see that commercialists are not ethically opposed to the adoption of corporate interests in their work. That is, in the moral reasoning concerning disinterestedness—whether it is good or bad, right or wrong for academic scientists to engage in research that is motivated by external, not "purely intellectual," interests—commercialists support departure from that norm. Having let go of disinterestedness, the primary issue for a scientist in negotiating the balance between academic and industrial interest becomes autonomy, the hallmark of professional control.

The key reason commercialists reject the notion of disinterestedness is because of the value they place on efficiency, another formally rational value that is central to the ideology of commercialism. Industrial reference groups offer efficiency and power. They have existing investments in specific societal problems and expansive resources in terms of funds for research, expensive analytic instrumentation, and a reserve army of industrial scientists to perform the routinized aspects of research. Industrial reference groups thus enable scientists to wield considerably more power than would be possible exclusively within an academic laboratory. We can see the emphasis placed on efficiency in the comments of "Roger,"[12] a biologist who leads a corporate-funded research institute at a public university. As Roger and I discussed industrially sponsored research he commented:

> If you want to make change, one of the ways of making change is to grab one of the big energy companies and help them do the right thing. From my inter-

> actions with [an energy company], I became convinced that there was a lot of will inside the company to try and change how they were making energy. They wanted to evolve the company. And so I felt I could help them do that. The resources they were putting into it were enough to make a significant impact, and so that's why I [backed out of my startups and] decided to come.

In Roger we observe a scientist who divested himself of four of his companies to adopt and influence the goals of his industrial sponsor. In his view, the power he can wield through a corporation—in this case, one of the world's largest energy companies—far exceeds the resources at his disposal as an academic scientist with no external collaborations. In this respect, we see again the operation of predictability as a core element of commercialism in that corporations are committed to specific problems and mobilize resources toward their resolution. Commercialism can thus be viewed as a response to incentives that encourage specific paths of research.

Traditionalists generally reject external reference groups and seek to influence and be rewarded by groups within the scientific community. This in part stems from a different conception of utility. Whereas commercialists align with external reference groups like corporations because they perceive such alliances to efficiently foster scientific progress through the provision of financial resources, traditionalist conceptions of utility are anchored in impact on the scientific community. As a professor of biology known for critical discoveries in genetics explained to me,[13] "I think of utility in terms of the effect on the field. I think you want to do something that isn't totally obscure."

Some traditionalists do not view industrially sponsored research as inherently problematic. That is, they do not perceive industrial sponsorship as a means to exert undue influence on scientific discovery. The issue for these scientists, however, is whether or not funding is predicated on conditions, or the attachment of "strings." I spoke with a professor of biology at a private university who,[14] when discussing his views on sponsorship from corporations, explained:

> You can't be doing work for service. That's completely unacceptable. If a company needs a particular type of experiment done and you happen to know how to do that experiment, you should not agree to be paid to do that experiment. I think it's fine to have people from the company come and say, "Show us how to do this." But you should not do fee-for-service. I just don't think that's appropriate.

His account can be contrasted with that of the commercialist chemist who does not regard such arrangements as "ethically abhorrent," a comparison that reveals that commercialists and traditionalists disagree about the appropriateness of outside groups influencing academic research. Traditionalists believe that academic scientists should be undirected by influences outside of the university. Disinterestedness is operative in the sense that the criteria by which scientists select their problems originate from scientific, not external, concerns. Yet as the following account of a traditionalist associate professor of chemical engineering illustrates,[15] unrestricted industrial sponsorship may nevertheless undermine the operation of disinterestedness.

> The problem could be if, in pursuit of corporate money, I spend a lot of time just coming up with demonstrations. So the amount of time I spend on basic research, that is squeezed, and I spend more time at making the work flashy or attractive to these companies in the hope that these companies will then give me money. That is going to compromise the integrity of my research program. It's going to take me away from what I really ought to be doing, and it might even cause me to rear away from the bigger questions in pursuit of specific questions that I think might be of interest to the company.

Traditionalists believe that commercial incentives, whether symbolic or material, divert a scientist from the problems defined by authorities within their field as most critical to scientific progress. From this perspective, even unrestricted funds are seen as a threat to science because of their source. Commercially based unrestricted funding is viewed as a soft form of influence that leads scientists to shift their interests away from the frontier of research and toward areas of science associated with the concerns of a company. Thus, according to traditionalists, industrial sponsorship constrains disinterestedness through the introduction of nonscientific incentives and alternative audiences for one's work. Such a perspective scarcely acknowledges the possibility that an overlap could exist between industrial concerns and the concerns of the research frontier.

Universalism

Universalism denotes a mode of evaluation: are scientists and scientific research to be judged by a universal frame of reference or by particular attributes? Universalism as a norm holds that scientific merit and the quality of role performance serve as the only basis for decisions regarding appointments, promotions, fellowships, honors and other rewards, or resources for research.[16] Research on universalism in science examines how particular social attributes, such as institutional affiliation, gender, and race, interfere with the allocation of awards based on scientific merit alone.[17] Our concern here is to consider the relationship between commercialization and universalism, with the objective of understanding both scientists' perspectives on commercialization as a basis of merit for rewards and also the rationales that motivate these views.

With few exceptions, all universities claim the intellectual property that underlies any potential royalties derived from commercialization and permit scientists a share of such royalties that result from their inventions. Universities therefore do reward scientists for commercialization in this way, but they also engage in other formal and informal processes of allocating rewards to scientists, including the distribution of laboratory space, allocation of teaching requirements, provision of resources for research, and decisions related to promotion and tenure. Although the majority of the scientists in the study believed that commercialists legitimately deserve royalties resulting from successful commercialization, the views of commercialists and traditionalists diverge with respect to whether commercial success should factor into promotion and tenure decisions.

Views of Commercial Rewards: The Market as Peer Review

Successful technological and market-related endeavors are central to commercialists' definitions of success and professional identity. Given this centrality, it is unsurprising that these scientists believe their commercial activities should be incorporated into decisions regarding their promotion and tenure. Commercialists I interviewed emphasized that although commercialization should not be a stipulation for role performance in academic science, it is a form of scientific output that should be considered when evaluating the work of a scientist. When I asked a professor of biochemistry in her forties whether universities should reward scientists for commercializing their work,[18] she responded that achievements such as her startup company "certainly should go in the package." She explained further: "Young assistant professors should get something, in addition to just publication credit and committee credit. They should get some credit for patenting and commercialization activities."

There are a number of reasons commercialists believe that market success should be factored into promotion and tenure considerations. Because commercial practices are associated with scientific work, commercialists view them as a component worthy of consideration equivalent to consulting, public service, or other activities in which scientists are engaged. Commercialists also believe they should be rewarded because their work has contributed to their own visibility or that of their department or university. Rose,[19] a commercialist professor of chemical engineering in her forties we have already heard from, explained:

> It's meaningful in our greater community to recognize when there's a wonderful industry-academic collaboration. People see that these academicians and these industrial scientists got together and they created a world of knowledge which has become central to both manufacturing something and some critical understanding of polymer science. Well, in that case, wow! That's one of those "wow" things, and we all appreciate it. And when that person is being evaluated for something like tenure, those kinds of awards are important to hear.

A fifty-year-old professor of chemistry similarly stated:[20]

> Should they be rewarded beyond [royalties] with pay raises and things like that? What is the goal of a research university? I often think that if fame is part of it, we are rewarded for fame, and so to the extent that [commercialization] gains fame for someone in the university there should be some additional reward there.

Visibility has always played a role in the scientific reward system. Only discoveries that come to be known can influence the advance of knowledge, and therefore the appropriate allocation of rewards is dependent upon the effective communication of ideas. Visibility is gained through the publication of high-quality work, but a scientist's visibility also benefits from honorific awards, departmental prestige, specialty, age, and collaborators. Commercialization, in the eyes of its practitioners, constitutes yet another means by which scientists can enhance their visibility, which is of benefit to them and their department. Is it peculiar that commercialists believe they should be rewarded for increased visibility? After all, visibility is an effect of scientific achievement—a second-order effect of commercialization—but not necessarily an act that itself contributes to the advance of knowledge or technological development. Commercialists see visibility as legitimate grounds for professional reward because they regard the market as the ultimate form of peer review. As a professor of chemical engineering explained as we discussed whether commercialization should factor into promotion considerations:[21]

> If an investigator has patents that have been licensed and commercialized, then there is a peer review on the quality of that intellectual property and it's the marketplace. So I think factoring that into the impact assessment is a legitimate thing to do.

This idea ties back to a pattern we considered with regard to research problem selection in which commercialists view utility as enriching the quality of scientific research: commercial success is viewed as an indicator of quality because it calculates the level of utility present in the commercialist's invention. Commercialists thus support a reward system in which judgment by a nonscientific audience about the value of commercial practices factors into organizational distribution of rewards. In sum, commercialists believe they should be rewarded for their commercial endeavors both monetarily and in terms of promotion and tenure. They see commercial success in the market as institutionally relevant: a rationale that constructs commercial activities as legitimate, esteemed, and beneficial to university and departmental prestige.

Universities have an inherent interest in rewarding commercialists because a university prospers when its commercialists are successful. But my interviews suggest that how universities reward commercialists is often controversial because doing so can contribute to inequality through its impact on the autonomy of scientists.[22] In a biology department at one of the public universities in the study, for example, the preservation of a stream of income generated by one scientist I interviewed, "Campbell,"[23] conditioned the administration's choice in hiring another scientist. Campbell has generated several million dollars in university revenue through licensing of his patents. With a nine-month academic salary that exceeds $200,000, he is among the highest-paid scientists in the study (and at his university). During our interview, he emphasized that universities should not reward scientists for their commercial activities:

> Universities should reward the intellectual aspects. That's what we're hired for. The commercialization we end up benefiting from anyway. So I think that would start degrading what an academic institution is all about if we say . . . "[Campbell"] has hundreds of patents, he should have the highest salary in the university." That's not what we're about. We're about teaching, we're about doing basic research, we're about finding out new things, and that's how the university should judge us.

In contrast to Campbell's own view, other scientists in his department perceive that their university does indeed consider commercialization an important factor in how organizational rewards are allocated. A colleague was discussing how his university rewards commercialists when he told me the following story:[24]

> Let me just give you an example that illustrates that there is not 100 percent consensus about [commercialization]. You know [Campbell] is the [technology]

> guy here. He brings in a lot of money. The tech transfer office loves him, and they take care of him. He is not young anymore, so a few years ago the tech transfer office decided that in case [he] were to croak off, we needed to have an expert in place here who knew [his technology]. The technology transfer office actually put up the money for us to hire a new professor, but the string attached to that money was that it had to be somebody who worked with [Campbell's technology]. People didn't like that. They didn't like the idea that the department would let its future direction be dictated by commercial needs. The new faculty member didn't exactly get a resounding majority. We hired him, but it was not something we did cheerfully or wholeheartedly.

This example departs from the universalistic ideal, which includes the notion that a hiring decision should be based on preestablished intellectual criteria, not on a candidate's ability to sustain the university's investment in a specific technological area. The case illustrates the restriction of careers on grounds other than competence because the economic interests of the university took precedence over general scientific merit.

Views of Commercial Rewards: The Academic Caste System

Overall, traditionalists are concerned about how scientists are rewarded for commercial endeavors, fearing that universities' incentives to reward commercialization will create unequal conditions among scientists and distract scientists from basic research. A minority of the traditionalists I interviewed even felt that commercialists should receive no material rewards whatsoever for their inventions. As a professor of biology explained as we discussed how scientists should be rewarded:[25]

> I think basically things that NIH funds that end up having amazing commercial implications—that money ought to come back to the government. I don't think that's legitimate to then go off and own the company. I think . . . the price that scientists ought to pay to have NIH support is that then the taxpayers, whose money that is, get a big chunk of whatever benefit there is from that. Not just the item, but the money from the item.

This perspective is rooted in concern about not only the ethics of professional rewards but also government funding for scientific research. Some traditionalists perceive that federal agencies, which provide funding and thus enable individual and university royalties, are underresourced but receive no material return on their investment in research as ethically problematic. Many traditionalists, particularly those at the public universities I studied, depict themselves as "public servants" because their research is funded by taxpayer money, which could be an unstated reason that some perceive the lack of commercial returns to federal agencies to be an issue. Yet none of the traditionalists who critiqued the private exploitation of federal funding provided an alternative model to the economics underlying commercialization.

Most of the traditionalists I interviewed embraced the economics of the current technology transfer regime, but they were concerned about how rewards for commercialization are allocated within universities. They generally opposed what they believed would be an inequitable distribution of rewards for role performance. This view focuses primarily on potential differences in the conditions of work for commercialists and for traditionalists. Questioning the impact of organizational rewards for commercialization on equality, a professor of mechanical engineering in her early fifties asked:[26]

> What I'm struggling with is: If they're rewarding people for [commercialization], are they not rewarding people for other things? Does it create a class or caste system? That would be my worry, that people are seeing that that's where the brownie points are, and so is there a push to commercialize more things.

Although she did not specify how such a "caste system" might operate in academic science, other traditionalists I interviewed did describe ways their universities reward commercialists that can be detrimental to traditionalists. "Ruth,"[27] a professor who completed her PhD in the 1970s, depicts how commercialists "write their own ticket," which she says is "absolutely common."

> The universities get a kickback from the professors who have their companies. Those guys buy us laboratories and stuff, and we pay for it in ways that are subtle. For example, some person who has an important industrial tie and who would never consider resigning his university job because he likes the prestige of being a university professor gives us a bunch of money, and in return, he gets to teach whatever he wants. And all the hard work gets done by the peons, the assistant professors who have to teach the terrible courses. That's absolutely common in our department, and probably in other departments, too. The tycoons write their own ticket when it comes to the jobs they do in the department.

Traditionalists recognize that their universities benefit monetarily from the successful commercial activities of their scientists, but they reject the notion that commercialists should receive special resources that encourage commercial success. The finite nature of laboratory space, money, and faculty means that support for one scientist's obligations may come at the expense of his or her peers. Such a scenario represents the threat that some traditionalists fear and reject: indirect "punishment" through positive discrimination. Commercialists benefit from activities that traditionalists forego or avoid, at the expense of traditionalists.

Traditionalists also oppose the university's allocation of rewards for commercialization because they perceive it as an obstruction of "pure" scientific progress. The university, as the organizational apparatus through which material rewards are distributed, influences the scientific reward system through its impact on the opportunity structure in science. The recognition that motivates a career may come from the scientific community, but universities control the environments in which careers are carried out. As a consequence, universities can influence the advancement of knowledge by incentivizing particular activities or fields, potentially at the expense of others. To examine this point, consider my interview with an associate professor of chemistry in his late thirties.[28] He told me he did not have a "clear answer" regarding whether or not universities should reward scientists for commercialization, and then he continued:

> If you have to push me to one, then I would say probably not. Somebody's research, patents, and so on . . . they benefit from that. [There is a] licensing fee that can benefit their research or even their personal life. That is perfectly fine. But I don't think that commercialization should be part of promotion considerations, because then one is forced to question, What is more important: fundamental research or commercialization?

From the perspective of traditionalists, fundamental research is the chief priority of the university. To provide incentives for commercialization, they claim, constitutes an undue threat to this objective. The traditionalists I interviewed did exhibit a permissive view of commercialization, but such tolerance was contingent upon commercialization's status as an adjunct activity of, not a motivation for, science. According to this view, commercialization is a legitimate ancillary activity of fundamental research, but it should not be the institutional goal because organizational rewards for commercialization could threaten the allocation of effort toward achieving science's true goal, which is the pursuit of knowledge. This view of rewards, motivation, and scientific progress is evident in the account of a professor of biology who envisions how commercial rewards could influence his scientific commitments.[29]

> I think it's fine if the rewards are rewards of the sort that stimulate fundamental research. If the money I make on a patent were put not into my pocket but into my lab, that's fine because I could use that money to do fundamental research. If the reward encourages me to spend more time with a company or to be a little more inclined to have a student work on a project relating to the company, that is a corrupting influence. Science should be pure in the sense that your only concern is to answer a fundamental question. If you stumble across something useful, all so much the better, but if you're always keeping your eye on the bottom line, you're not behaving the way a really good scientist behaves. Really good science is done by people who focus exclusively on it, and everything that distracts you from that is likely to make you do less good science. I feel that very strongly.

According to this view, science and commercialization are opposite goals that should not compete. In other words, commercialization is an institutionally irrelevant rationale for receiving professional rewards.

What are the effects of commercialization on the operation of a universalistic reward system? The interviews suggest that commercialization constrains the operation of the norm of universalism. Universities stand to gain financially from the commercial outcomes of their scientists, a fact that creates an organizational incentive to develop environments that encourage commercial pursuits. This could mean alleviating commercially successful scientists of burdens such as committee work or undergraduate coursework, or favoring appointments of scientists with commercially relevant agendas over traditionalist researchers. Commercialism could thus function as an achieved status, which, like institutional affiliation and doctoral origins, could foster particularism in science.

If universalistic standards of evaluation are reduced or disappear in an era of commercialized science, how might scientists themselves be affected? One possibility suggested by the interview data is that commercialization could function as a new basis for the accumulation of advantage: hiring procedures could favor doctorates from commercialist laboratories or institutions; resources for research could be unevenly distributed among scientists in ways that favor commercialists; or commercial royalties could be circulated primarily to other commercially relevant fields of research. Commercialists and traditionalists could therefore potentially pursue careers in fundamentally different opportunity structures, situated in higher and lower prestige stratas, respectively.

Sociologists Jonathan Cole and Stephen Cole raise a question about science quite relevant to this discussion: are criteria of evaluation "which are irrelevant from the point of view of the individual scientist, very much relevant to the institutional goals of the system?"[30] In other words, can discriminating based on commercialization possibly benefit science? Commercialists believe they should be rewarded for their commercial successes in addition to their scientific competence because such activities bring money and societal support to the institutional goal of science. Traditionalists, by contrast, believe that such benefits are at best short term and that rewards for commercialization distort the direction of science toward its goal.

Communalism

Among the three scientific norms that surfaced in my interviews with scientists, communalism is the most directly tied to the reward system in terms of individual scientists' behavior.[31] In order to achieve priority in discovery, scientists must share their findings, for without sharing one's results it is impossible to achieve any credit or recognition for one's work. Communalism thus aligns individual and institutional goals by basing the allocation of rewards upon sharing findings with one's peers.

Communalism merits close attention in analyses of commercialization. The notion that knowledge is "owned," or the property of any entity and therefore to some degree secret, is contradictory to the tenet of communalism that declares knowledge to be the product of social collaboration and therefore public and meant to be shared. Given this conceptual tension, it is no surprise that some scholars believe commercialization may undermine the norm of communalism in the scientific community.

Studying secrecy in science is akin to the study of scientific fraud, as practitioners of fraud are unlikely to willingly reveal their indiscretions. Moreover, the university is highly protective of any incidents that could potentially discredit one of its scientists or tarnish its own reputation. When scientists disclosed their observations of concrete acts of secrecy, they did so coupled with appeals to the anonymous nature of the interview. Some scientists, including some who witnessed inappropriate secretive behavior while serving on conflict-of-interest committees, declined to disclose details. In this section, therefore, I organize the discussion around the types of constraints on communalism that emerged in the interviews, given that instances of observed secrecy were reported by both commercialists and traditionalists. The interviews uncovered four contexts in which secrecy undermines the operation of communalism: peer reviews, conference presentations, graduate training, and publications (see table 5).

Poaching and Sandbagging in Peer Review

The review process of research and of proposals for funding, in particular, constitutes one of the earliest contexts in which new ideas and findings are presented to one's peers.[32] Many of the scientists I interviewed emphasized that these reviews present an opportunity for reviewers to poach ideas or to undermine threats to their own agenda by sandbagging, which refers to blocking the progress of other scientists by rejecting proposals that compete with their own research agendas. As the boundaries between science and commerce have been rendered more permeable, these practices may also be driven by commercial concerns. Consider the following story, shared by a commercialist professor of chemistry in his sixties,[33] as we discussed his perception of problems that arise from commercialization:

> I got a review from the NIH where an anonymous reviewer wrote, "Well, this is all really interesting work, nice proposal, but I happen to know that there's a company that I can't name that is working along similar lines, and they're pretty far along with their technology and it really is just as good as what these people are proposing. Therefore, this grant should not be funded." So, unnamed company, anonymous confidential information, unnamed reviewer—clear conflict of interest because the person is consulting with the company or else they wouldn't have the information. They used that information to sandbag a proposal, and this is accepted by the NIH reviews. Now when that happens, you know, you've gone too far, frankly.

One interpretation of this chemist's account is that the norm of universalism was undermined because nonscientific reasons—namely, the existence of competing yet unpublished work not associated with this chemist—factored into the evaluation of his work. Another is that the rejection may have been a legitimate exercise in "efficiency," so this scientist could avoid repeating work already carried out by other scientists. However, the most plausible interpretation, given the chemist's emphasis, is that secretive information—in this case, the reviewer's access to proprietary knowledge—interfered with the proposal being reviewed according to exclusively scientific criteria.

Posturing and Positioning at Conferences

A second context in which scientists suggest that commercialization undermines communalism is conferences and seminars. In this context, secretive behaviors include posturing and positioning. Scientific meetings provide a context in which ideas and discoveries are shared prior to formal peer review or certification as accepted knowledge. Secretive behavior by commercialists at departmental seminars was mentioned by a traditionalist associate professor of chemical engineering in his early fifties,[34] who explained to me:

> When people are playing their cards close to the chest or when they're being proprietary about something, I see that as being antithetical to the pursuit of knowledge. When I see some seminar speaker from academia who says, "I'm doing this, I'm doing that, we've got some really nice results, but I can't talk about them yet because we're patenting them, or because it hasn't been released by the sponsor," I always think: why are you bothering to waste our time with this? Talk about it when you can talk about it. But when people get into being coy about commercializing something, I feel like they're trying to have it both ways. They're trying to say "I've done it, but I'm not going to tell you about it." . . . A few of my colleagues are impressed by that and get all psyched up and eager. I'm put off by it. There are negative impacts, in that it leads to some positioning, staking of territory, and it's just bad behavior. It slows things down if they're making a big deal out of it but not saying anything about it . . . They're trying to scare away the competition.

Science is at once communal and competitive. Scientists compete for priority rights in discovery even as they draw upon the collective work of the community and share the results of their work so they can be replicated. To be sure, scientists I interviewed pointed out that it is not uncommon to hold discoveries "close to the chest," referring to the tendency to delay publication until results have been verified. However, the chemical engineer's perception in the preceding account suggests something more: posturing to protect one's position in the field by communicating to one's colleagues: "Don't waste your time, you're behind me," and "bluffing" by referring to unverifiable and secretive results in an attempt to preliminarily assert one's claim to priority of discovery in some area of research.

Restricting, Rerouting, and Misappropriating the Work of Graduate Students

In their work on the normative structure of graduate education, John Braxton, Eve Proper, and Alan Bayer identified inviolable norms that provoke indignation, such as harassment of students, and admonitory norms, such neglectful teaching, that faculty view as mildly inappropriate.[35] Here we peer into proscribed behaviors in commercial contexts that arise when secrecy interferes with communalism in graduate training. Threats to communalism may be subtle, informal, and occur in more than one way. The simplest way is restricting the ability of a student to publish his or her research. During my interview with a commercialist professor of chemistry[36] who was in the process of starting his first company, he described how the industrial sponsorship at his university resulted in restriction of graduate student research:

> Can the student describe any of that [industrially sponsored] research in a thesis? If it's really meant for commercial purposes, the answer to that would be, of course, no. Or, yes, but it would take two years, and the student has to work in sort of this guarded way. That is not appropriate at all. There are plenty of people who do [this].

The professor's characterization of the practice of utilizing doctoral students in commercially oriented projects as commonplace is noteworthy because policies at the universities I studied prohibit it. My interviews suggest that involvement of graduate students in commercially oriented work is not uncommon, and one scientist suggested that commercialists "push the boundaries of student involvement" in commercially oriented work.[37] Ruth,[38] situated in the same department as the chemist just quoted, corroborated his point:

> I have seen people who had their graduate students actually doing research at their companies, which is strictly against the rules and would have been a firing offense if it had been widely known. The students in question were severely conflicted and had a lot of problems. I'm talking about something that's really confidential, because I hear about it in ways that aren't public channels. People end up doing things that are injurious to their students, and that's the worst thing you can do.

While secrecy is detrimental to the progress of science, its consequences for individuals vary according to where one is in one's career. The consequences of delaying publication are highest at the beginning of a career, because publications are critical to attainment of an academic appointment. Over time, particularly after achieving tenure, the sanctions for secrecy progressively diminish. Thus, the individual consequences for violating communalism are the highest for doctoral candidates whose opportunities to publish research are restricted or even misappropriated by their research advisors.

What is more, the stakes vary according to how proprietary research is organized. The least deleterious circumstances are sponsored research agreements with major companies. Large corporations are able to patent quickly because they possess extensive legal resources to protect their technologies. Thus, scientists and graduate students involved in such research are able to freely discuss related research with peers, at meetings, and in scientific publications. Startup companies founded by professors, by contrast, are not as well equipped. With fewer financial and legal resources at the disposal of such companies, there is increased incentive for scientists to hold on to trade secrets. The problem this presents for both commercialists and their doctoral students is evident in a story that a commercialist professor of chemical engineering told me about a company he founded and why he left it:[39]

> A couple of years ago, I left [my startup company] completely because I saw a major conflict of interest coming . . . A small company can take the attitude that they don't want the patents because the patent becomes public, and if a big company wants to take it, they just can infringe on it and influence the court and throw a thousand lawyers after one poor lawyer for the startup, so there is something to say for a small company doing trade secrets, keeping everything secret. That means that they don't want your students knowing what's going on because they're going to graduate and they're going to go to a competitor and they're going to tell them trade secrets that aren't patented, so now this whole issue with secrecy becomes a minefield because you don't have the patenting which then you can talk about afterward. It's just indefinite trade secret. I could see the whole net and the whole web out there, and before it got dicey, I just completely cut off relations with [the company].

Apart from the "minefield" of secrecy that academic startup companies must navigate, and the climate of confidentiality they introduce within universities, this scenario suggests that there is a commercial incentive for scientists to avoid including their graduate students in commercial work involving unpatented trade secrets, since after graduation their knowledge could end up in the hands of competitors. How commercialists navigate decisions regarding the involvement of graduate students in commercially oriented research is rarely free of complexity, especially in the context of company formation. A commercialist associate professor of electrical engineering I interviewed,[40] who started his first company as a graduate student, described these decisions as "slippery slope" situations:

> Let's say I want to start a company and I see that there are certain things I need to put in place first. I need to do these experiments. So I say, okay, fine. I have some unrestricted funds. I have these students working on these projects that

> are sort of related. I'll have them do it. So I know a little bit. It's a pure scientific idea. A student works on it for a while, but it starts to coalesce in my mind that it is actually a company now. Where do I stop the students? When do I say: you can't work on it anymore? They might have worked on it for four years. Should I stop them from working on that now, [when it is moving] in the most profitable or most exciting direction, because it actually leads to a real impact? Or instead do I sort of divert them to sort of a less fruitful avenue of investigation? Those are real dilemmas, right? Where do you draw the line? This is a slippery slope type of a situation. I started out with a great scientific idea. I ended up with a company.

Secrecy may impact on graduate students in at least three ways. One way, as we have seen, is restriction, in that students' ability to publish certain research may be restricted. Their ability to communicate with their advisor or other scientists may likewise be constrained by secrecy associated with the laboratory in which they work. Their ability to pursue problems of interest—such as those related to projects that lead to commercial outcomes—may also be restricted due to the proprietary interests of a firm or their advisor. In this scenario, an advisor's interest in commercializing may lead him or her to reroute or divert a student away from commercially profitable intellectual property. Finally, in the worst scenario, is misappropriation: the ambiguity of distinguishing between publishable and proprietary research may shape a scientist's willingness to assign credit to the student for intellectual property, and for the student to recognize it himself or herself. While misappropriation and restriction are viewed as inviolable or clearly inappropriate, rerouting may generally be viewed as admonitory provided that it does not interfere with student progress.

Delaying and Withholding Publication

Secrecy undermines the operation of communalism in a fourth context: publication, where it can take the form of delaying or withholding publication or of suppressing the publication of negative results. In the scholarly literature on research policy and higher education, the suppression of negative results is framed as a conflict of interest.[41] The suppression of negative findings violates communalism because it cloaks the truth. There is little reward associated with negative findings. Journals rarely publish such research results, even though doing so could prevent other scientists from repeating the same failed effort. Such suppression constitutes a routine but potentially harmless departure from communalism. But when negative findings are suppressed to avoid undermining the market potential of commercialized research, communalism is clearly violated. When I asked commercialists and traditionalists about the suppression of negative research results, most of the scientists suggested that this type of problem is limited to clinical research at university medical schools.

Withholding research results and failing to communicate with one's peers was a core concern that emerged among traditionalists when I asked about the perception of detrimental consequences of commercialization. Few suggested that they had personally observed their departmental peers engaged in withholding, which after all is not easy to recognize, but a number of them did highlight examples they were aware of at their university. A traditionalist professor of chemistry at a public university,[42] for example, told me:

> I think we should always have a dialogue. We should always try to find each other. Of course that's a big problem [because] at the moment you have commercialization, you have secrecy, and there is now—now there is secrecy going [on] over there [referring to a biology building]. Everybody will deny it. But when I see that the students can't get the seminar room because the whole floor is off limits, then I say this is not right, okay? This is a university.

At each university in the study, policies prohibit restrictions on publication of research results under any circumstance. Furthermore, the universities permit periods ranging from thirty to ninety days to allow commercialists and their sponsors to review the potential intellectual properties disclosed in a publication. They may then file invention disclosures or provisional patents, if necessary. For commercialists to engage in secretive research or fail to publish findings due to an agreement with any company not only departs from the scientific norm of communalism, it violates the policies at their university.

Delaying publication represents the most common practice emphasized by commercialists themselves. The commercialists I interviewed provided numerous observations of ways that commercialization encourages secrecy in science, but when discussing their own practices, they invariably presented accounts of their work that were consistent with their university's policies. When I asked them about secrecy and withholding results in their own research, their responses generally mirrored the following account given by a commercialist associate professor of chemistry:[43]

> There is an element of secrecy, in that there's a couple of projects that I've discussed with my students that we're not going to tell the world about until we're

> ready to disclose it. But the intention is always that we will publish those and the students will get papers because they have to have that, right? So I think maybe it's more of a delay in disclosure than anything else.

In other words, commercialists framed their own withholding of results as a delay and therefore as temporary only. Nearly all of the commercialists I interviewed acknowledged that the process of research, invention disclosure, and the patent process results in some degree of secrecy and withholding. None suggested that withholding occurs more than a brief period of time.

One commercialist I interviewed, a professor of biological engineering,[44] indicated that commercialists are able to be secretive and communal at the same time. We were discussing the tension between sharing and secrecy when he said:

> I think there's merit to the tension and there is a trade-off. If there are trade secrets that are not fully disclosed because a company needs time to get the advantage that they need to get because of the risk they're taking, some of that's justified. I try to separate concepts from specific implementation sorts of things, so you wouldn't share with the whole world a specific code that you're using to help [a medical technology company] get a product out there based on your science, but you shouldn't hesitate to call your colleague at the University of Utah who is using the equipment [of the medical technology company's competitor] and talk about the concepts that are going on . . . It's not a completely black-and-white area. It has to be looked at carefully, I think.

Commercialists see a place for both secrecy and openness in science. This example and others we have considered suggest that communalism is destabilized by commercial practices in science.

Conclusion

The norms that scientists espouse are a key dimension of professional ideology in the academic profession because they reflect scientists' beliefs about how faculty and universities should work. The ideology of traditionalism in science—and its imperatives, such as conducting science for the sake of science—remains important to many faculty. This is a much different pattern than implied by the prevailing transformation hypothesis proposed by some scholars, which suggests that Mertonian norms are no longer relevant. Nevertheless, traditionalism bears an increasingly tenuous influence on science at elite universities as the ideology of commercialism, driven by a new institutional goal of creating technologies that will have a societal impact, is incentivized, embraced, and rewarded. Commercialists espouse commitment to formally rational elements of efficiency and calculability. They seek efficiency through partnerships with industrial reference groups, enabling them to wield considerable power through access to deep financial, human, and technological resources. Commitment to calculability can be seen in their emphasis on the pursuit of concrete returns on investments in science and their embrace of the market as the ultimate form of peer review in higher education. Commercialism has a contested legitimacy among traditionalists, while traditional norms are deemphasized among commercialists, indicating the presence of competing codes of conduct among elite scientists alongside competing visions of what scientists at universities should do.

Our understanding of professional ideologies in science is incomplete, however, if we only examine faculty perceptions of the role of research in higher education. Professional ideology is also embedded in the status systems that faculty construct around rewards for roles associated with each view of the university. To complete the picture of ideologies of science in an era of academic capitalism, we must therefore turn to meanings and status systems associated with commercialism and traditionalism.

The Reconstruction of Meaning and Status in Science

In addition to the "oughts" of how academic research should operate that are associated with commercialism and traditionalism, professional ideologies also specify which roles are most highly valued. We therefore need to hear how scientists construct meaning and status in academic capitalist contexts—how they conceive their roles, what they esteem, and the activities they shun—to achieve a more complete perspective of commercialism and traditionalism as professional ideologies of academic work. Accordingly, the purpose of this chapter is to understand the implications of a commercial reward system for meaning and status in academe.

I examine meaning and status systems as moral orders. Moral orders are collectively shared views that articulate determinates of prestige and excellence, the tasks that are considered the most difficult and valued, and what it means to be exceptional. Status systems within professions are contested and differentially valued, and in an environment where everyone has federal funding and is publishing in journals such as *Nature* and *Science,* commercialism provides a new basis for distinction. But competing reward systems have severed the meaning of status in science, revealing a new pattern of division and conflict related to how scientists construct and define achievement.

The dimensions of the traditionalist and commercialist moral orders of science I uncover coalesce into two general modal patterns that characterize the nature of scientific work within each. Moral order falls into two general patterns among traditionalists and commercialists. Traditionalists' views of science and the practices in which they prefer to engage are governed by professional criteria only. Thus status within the ideology of traditionalism follows Abbott's purity thesis; that is, it is a function of professional purity, or the ability to work exclusively within the knowledge base of science—free from applications, corporate interests, or any other concern that "taints" the elegance of pure science.[1] "All we want you to do is change the way we think about science" a senior scientist told a newly appointed chemist I interviewed[2]—a grand challenge that parsimoniously captures the logic of professional purity. But Abbott's purity thesis, and a similar argument made by Merton, does not hold for the ideology of commercialism, in which the moral order of science, by contrast, is characterized by professional rebellion.[3] Commercialists reconstruct the goals of science to reflect fidelity to societal rather than professional problems, to embrace hierarchical modes of organization around this end, and to construct standards of eminence and success that depart from the community of academic scientists. Merton referred to such patterns of adaptive behavior as rebellion,[4] yet he never postulated that such patterns might constitute a basis for eminence in science.

Through the interviews I conducted with scientists across the United States, I discovered that professional purity and professional rebellion, the moral orders of traditionalism and commercialism, are embodied in four dimensions of scientific work: the logic and organization of scientific work, the status scientists seek, the reputations they attempt to construct through visibility, and the activities they avoid, as shown in table 6. Traditionalists' quest for status through purely scientific means shapes all aspects of their work. They organize their research groups collegially, using a craft-like group structure predicated on devotion to graduate training, which they view as one of the key forms of impact on the scientific community. They aspire to be known primarily for changing the way others think through contributions to knowledge. Consequently, they avoid activities far removed from the knowledge base of their field, such as bureaucratic tasks or commercial applications that do not substantively contribute to the reputations they desire. Commercialists rebel from these traditional elements of a scientific career, drawing instead on such formally rational elements as control, efficiency, and calculability in all aspects of their work. Motivated by a desire to be known for

TABLE 6

Moral Orders of Commercialist and Traditionalist Science, by Dimensions of Work

	Traditionalists	Commercialists
Organization of work	Collegial	Hierarchical
Bases of status		
Income	Average / Above average	Above average
Predominant professional product	Scientific knowledge	Corporate technology
Visibility sought	Scientific impact	Societal impact
Dirty work designations	Administration	Administration
	Directed research	Directed research
	Incremental publication	Incremental publication
	Commercialization	Book publication
Overall modal pattern	Professional purity	Professional rebellion

creating technologies that solve societal problems, they organize their research groups hierarchically with the objective of facilitating both academic and commercial work. Commercialists eschew activities perceived to lack or have only minimal societal impact.

The Organization of Work

To understand the nature and meaning of scientific work, we must understand how it is organized. Our concern regarding the organization of work is limited to the structural properties of commercialist and traditionalist labs and the corresponding logics commercialists and traditionalists assign to how work is organized in their research groups. Roger,[5] whom we heard from in the previous chapter, summed up the two views from a commercialist point of view. I interviewed him at the corporate-funded research institute he currently leads at a public university. Prior to assuming this role, he had founded four companies, become a full professor at an elite university, and established himself as a leader in his field. As we talked about why he left his university to lead the institute, Roger explained:

> To get tenure at ["Chicago"] twenty people have to say you're in the top five in the world in your topic, so there's no doubt you have to be good at it. But what you actually do they don't care. And that's a huge weakness of universities because it's just random shots academically—little random bits of knowledge into the ether. If you suddenly have a big societal problem, where the hell are you supposed to turn to get a consolidated response to the problem? Who exactly is supposed to solve big problems that require large numbers of people working together? It's not clear that we have a system that really addresses that. So what we're actually doing at [this institute] is to get faculty working together to make a coherent whole so we can address big questions. I have still a research group, but I spend a lot of my time meeting with faculty, bringing together people with complementary expertise and getting them to work together as teams to go after bigger problems than they ever would as a university professor. And people like it because they can see they're contributing to something bigger than they could ever do as a professor.

Roger's comments are useful for identifying the distinctive meanings assigned to work in commercialism and traditionalism in that his narrative reveals distinctive logics of research, work organization, and goals. As he sees it, traditionalist scientists who eschew commercialization, even those at premier universities who are considered among the scientific elite, produce random knowledge of limited relevance to "big" problems in an uncoordinated fashion. Commercialists, by contrast, make contributions that seek to control important problems of uncertainty in society and thereby supersede the capacity of traditional scientists. From their perspective, the actions of scientists should be rationally organized around clearly defined and targeted societal problems. Here we encounter other formally rational elements of commercialism—control and predictability—in that commercialists derive meaning from doing work that seeks to eliminate uncertainties in areas such as human health or renewable energy.

Indeed, critical to this logic of scientific work is a division of labor predicated on hierarchy, coordination, and large groups comprised of experienced rather than aspirant scientists. In the work of classical sociologist Max Weber, this mode of hierarchical organization is intertwined with rationalization, a process in which traditional and value-based motivations for behavior in society are replaced by rational and calculated motivations.[6] Historically, science, like other professions, has resisted rationalization and bureaucratization processes, but here we observe how a segment of the scientific community seeks to wield new modes of power by embracing such formally rational values and processes tied to a mandate for societal impact.

This view and logic of work is not limited to the commercialists in the study who lead research centers or institutes at their universities. It can be applied to department-based research groups as well. Consider table 7, which presents ranges and average research group size of commercialist and traditionalist chemists. I have delimited the focus to chemists to control for differences across disciplines in the size of research groups. Commercialists operate substantially larger research groups than traditionalists do. Commercialists oversee about eighteen

TABLE 7
Research Group Size among Chemists

	Commercialists	Traditionalists
Graduate students		
Average	12.8	6.8
Minimum	6	2
Maximum	20	12
Postdoctoral scientists		
Average	7	2.8
Minimum	2	2
Maximum	18	4
Total group size		
Average	17.9	8.7
Minimum	14	2
Maximum	28	4

researchers, whereas traditionalists run groups of about nine.[7] One influence on group size is the number of postdoctoral researchers in commercialist labs. Traditionalists typically have two to four postdoctoral researchers, whereas commercialists have seven on average.

The differences between the two groups of chemists suggest distinctive modes of work organization. Traditionalists run what may be understood as collegial labs, in that a small group of researchers work collaboratively on a minimal number of grants. Commercialist labs, by contrast, are hierarchical, with postdoctoral scientists frequently making up half the research group, which allows greater efficiency in commercializing their work than would be possible otherwise. To understand the implications of these differences in group structure, let's consider the accounts told by the scientists themselves.

Commercialists: Hierarchies of Commerce

Commercialist scientists frequently oversee research centers and institutes in which commercialization is a central objective. These centers and institutes are sponsored by either federal funding agencies or corporations. The National Science Foundation, for example, funds centers that it claims provide "an environment in which academe and industry can collaborate in pursuing strategic advances in complex engineered systems and systems-level technologies that have the potential to spawn whole new industries or to radically transform the product lines, processing technologies, or service delivery methodologies of current industries."[8] One-third of the commercialists I interviewed currently hold or have held positions as directors in such units.

Even if a commercialist does not oversee his own research group and other faculty in an institute, governance of both academic and commercialist research operations in a company permit the scientist to approach work with the same coordination and view of "upstream" and "downstream" components of work. Or such coordination may occur within an academic lab, with the labor process divided according to what is considered appropriate for postdoctoral and graduate students. That is, in my interviews with commercialists, I learned that the presence of numerous postdoctoral scientists in their labs is often motivated by commercial objectives. Commercialists typically develop new technologies for which companies have limited expertise of the related fundamental science or limited interest in the necessary development that precedes commercialization. Since this work is usually not publishable apart from patent filings, it is considered inappropriate for graduate researchers, so it is assigned to postdoctoral scientists.

As I spoke with a professor of chemical engineering at a public university about his desire to commercialize research related to an article he had recently published in *Science*,[9] he spoke of "that intermediate ground where you have to do more applied stuff, not appropriate for the university, but not yet ready for the big company." He explained that his postdoc would "do this more practical developmental work and really see if there are any glitches in there that would preclude commercialization [and] get it ready to a point where now a big company might be interested in taking it over."

This division of labor becomes an instrument of boundary work, wherein scientists allocate commercial tasks to postdocs to keep graduate students out of the commercial realm. In this way, postdocs also facilitate efficiencies through management of different logics and rhythms of research. As I talked about the detriments of commercialization with a professor of chemistry who had founded an alternative energy company,[10] he emphasized the competing time frames of academe and industry:

> To publish a paper you've got to have a story, right? Sometimes companies want to see a quick answer. They don't want you to necessarily put together a complete story. And so from the standpoint of funding students, you have to balance that. And that's why when I get a grant from some place outside, my preference is to hire a postdoc, for example, rather than a student, because I think the postdocs are better able to handle that kind of, you know, a little bit of herky-jerkiness, that comes along a little bit.

The type of work that postdoctoral scientists conduct for commercialists is typically situated at the interface of academic and industrial science, rather than at the core of the knowledge base of a field. As a result, their presence in academic laboratories, apart from their own career objectives, enables commercialists to pursue both academic and commercialist agendas without violating acceptable standards of conduct for graduate training. Unable to allocate the tasks to graduate students because they are too routine and nonacademic, and unable to outsource the research to the market because firms lack sufficient fundamental knowledge, commercialists rely upon postdocs as "hired guns" to fill this niche.[11] Postdoctoral scientists therefore enable an efficient division of labor within commercialist laboratories, wherein different ranks of scientists are dedicated to different stages of the process by which fundamental discoveries are "translated" into commercializable products. This division of labor within a lab is particularly common among commercialists who have not formed companies but whose postdoctoral scientists account for half or more of their research group.

At the level of the organization of work, then, professional rebellion—and the ideology of commercialism more broadly—is predicated on a moral justification of control and efficiency. Commercialists argue that the pursuit of technological solutions to societal uncertainties requires departure from traditional modes of organizing scientific research, which are not conducive to efficient technology development because they conform to traditional norms of graduate training and do not support access to the expertise offered by a larger team of postdocs, faculty, and other technicians.

Traditionalists: Collegia of Devotion

In contrast to the commercialists' hierarchical mode of organizing work, the organization of work in traditionalist laboratories is collegial. By "collegial" I am referring to a guild-like mode of laboratory organization that involves less social distance between professors and graduate students than one finds in larger, hierarchical structures. Traditionalist research groups tend not to exceed ten members and are characterized by close collaboration and supervision. Although traditionalists recognize that large groups are indicators of success by virtue of research support, they see a size threshold beyond which the quality and coordination of research begins to deteriorate. The smaller size and greater supervision of research groups is one of the key ways traditionalists see themselves as different from their commercialist peers. A professor of chemistry who has trained almost forty doctoral students over the course of his career explained his preference for smaller groups:[12]

> I've typically had one or two grants, I've run a research group that has averaged five or six people, so I'm in that middle tier of people . . . I like to interact closely with my students, I spend a lot of time with them. I've had groups go over ten, and I've been unhappy with my management of it. You can't spend time with each individual. Say your week has left over after you've done your "must do" duties, fifteen to twenty hours a week to take care of your students. If you have twenty students, you've got one hour per week, per student. That's for everything: writing the thesis, writing their papers, discussing the research with them, developing their projects, managing them, whatever that takes, this is one hour a week for a human being, and I'm devoted to them during this time period.

His adherence to the professional purity is represented in his emphasis on devotion to his students, which expresses a high level of commitment to the institutional goal of training the next generation of scientists. But his comments also evoke the relationship between professional status and group size, implying that having a smaller group places him in the "middle tier" in his department.

Traditionalists juxtapose their view of how scientific work should be organized with their perceptions of what takes place in commercialist laboratories. Several of the traditionalists I interviewed perceive commercialist labs as hierarchical, and many were concerned about the impact of this on graduate training. Consider, for example, the view of a professor of chemistry at a public university.[13] When I asked him what distinguishes him from chemists in his department who have formed companies, he responded by comparing himself to one of his colleagues:

> This guy has . . . I know of a couple of [his] students that are graduate students. They were educated by postdocs. And now he doesn't hire hardly any students, and he has only postdocs and higher-ups because that makes the research run okay, and so he has given up essentially on the goal of what a university is because he does not train the graduate students any more. On the other hand, he's also director at [a research institute]. How can you do that? I have a group of twelve people. I'm working day and night on this! They have forty or fifty researchers there, he's got some companies, and he's on the board of some companies. How do you do that?

The perception that a peer has "given up . . . on the goal of what a university is" captures how purity and rebellion are each embodied in the organization of work. Merton considers rebellion either a rejection of institutional goals and the means to achieve them or the substitution of new means and ends. The professor's account suggests that the pursuit of commercial rewards requires a transformation of how academic work is done, such that what has been the core of scientific work—the training of future scientists—is deemphasized. Instead, traditionalists claim, the conditions of work for doctoral students in commercial research groups are characterized by the routinized patterns of work typical of industry. Ruth, a professor of chemistry from whom we heard in the previous chapter,[14] was particularly concerned about this influence of commercialization on graduate students' training:

> The work some of my [commercialist] colleagues have their graduate students do just astonishes me. They do things the way it's done in industry. You will have a bench chemist doing a reaction at fifty-two different temperatures and fifty-two different solvents, with no hypothesis, just trying everything in sight, completely mechanical. Where's the scholarship in that? That's a complete waste

of the student's time. [They do it] just because they do it that way in industry. There is no excuse for making a graduate student do that.

You have observed that this sort of activity takes place here at this university?

All the time. I don't understand it.

She later referred to this new organization of work as a "perversion of the educational enterprise," noting that it may have greater negative consequences for universities than for science. When I asked if there are ways it could hurt science, she said, "No, because it's not science." For her, such practices represent an abandonment of both university goals and science itself.

But the logics of purity and of rebellion with regard to lab organization are not just a matter of perceptions and preferences. They exist also in a broader environment of changes in how science is funded. One such change is a new emphasis on group-coordinated and technologically driven funding goals, the funding structure underlying centers and institutes supported by the government and industry. To understand how traditionalists regard this shift, let us consider the view of a professor of chemistry who received his PhD in the early 1970s.[15] As he and I were discussing the impact of commercialization on fundamental research he explained:

> I've had good funding from the beginning, but grants don't increase at the rate of academic inflation. So if a graduate student costs a grant $20,000 a year in 1980 and $66,000 a year now, my grants didn't follow. But I've been equally successful all along. I had a research group with nearly twenty graduate students. I was very intensely unhappy with that, and now my group is six students and one postdoc and I see it decreasing fairly rapidly, through no effort of mine. I'm doing the most exciting research of my career right now, and there is little chance that I can sustain a group of seven for more than another year or so. [This university] has created a harsh environment for people who do fundamental research because those people cannot sustain a career doing exclusively fundamental research. They have to get involved in practical stuff. They have to join group grants that are directed at fundable stuff. [Funding agencies] say we are going to create a new program focused on a particular problem, and to make this happen we won't work through the individual investigator direction, we will create large interdisciplinary programs where we get five to ten scientists working together with an obscene amount of money on a problem which we decide is important and we sell it to Congress. Then you have a five-year-ish program where these people have more money than they've ever seen before to do something which they may or may not be interested in and qualified

to do, but they do it and then they get used to that level of funding and they become parasites.

Here we see traditionalism constrained in an environment in which a purist orientation to science could also be interpreted as a failure to adapt. Whereas a parasite grows and feeds in the shelter of another organism, this chemist does not see a circumstance to thrive on his own terms. It is also evident that steadfast commitment to traditionalism may leave scientists embattled and with fewer resources for research. As we shall see shortly when we consider status, a traditionalist's refusal to adapt is a function of his or her perception of professional prestige, based on the degree to which he or she refuses to sacrifice his or her freedom in deference to outside demands.

Distinctions between the organization of work in traditionalist and commercialist research groups are thus shaped by the scientific goal (knowledge creation versus technology production), conventions of acceptable supervision and guidance of graduate students, and exigencies of funding. Traditionalists' preference for small or moderate-sized research groups is presented in light of perceived inadequate supervision of and insufficient contact with graduate students, and the strain they observe in large, commercialist groups. In contrast to the commercialists, whose work requires shielding doctoral students from involvement in intellectual property or commercial operations, the collegial organization of traditionalist research groups is predicated on the consistent overlapping of the goals of discovery and training. Commercialists, on the other hand, employ large, hierarchically organized groups with the goal of solving problems they view as "bigger" than traditional scientific work.

These divergent patterns of work organization suggest qualitatively different modes of scientific work between commercialists and traditionalists. Scholars of group processes have long considered group size critical to group behavior. Notably, research suggests that large teams are better at solving problems, but they are less creative than small teams. Sociologist of higher education Karen Louis, for instance, suggests that large research groups are best at "normal" science and less likely to generate novel ideas, whereas small research groups are more likely to contribute breakthroughs.[16]

This claim is consonant with the views of this study's participants. Traditionalists frame the objective of their work as making long-term, large, and transformational scientific breakthroughs and regard commercialists as focused on short-term, nonscientific problems. In fact, a brief examination of the vitae of the commercialists in this study suggests they have indeed made important scientific breakthroughs, even as they are committed to generating solutions to societal problems. What is less clear, however, is the approach to science that doctoral students who train in large commercialist research groups will develop. A comment by a commercialist in his forties highlights this concern,[17] and it supports the traditionalists' claim that research groups organized around commercial objectives inculcate a problem-solving rather than breakthrough-oriented mode of role performance.

> I think we're going to get a very different product of graduate student that comes out of research universities. While they may be more entrepreneurial, they may not be as deep scientifically. What kind of ramification is that going to have? It's hard to say. I think that they may have a nose for picking the kinds of problems that have great commercial impact. While nobody could deny that that would not necessarily be a bad thing as far as society is concerned, it could end up like combinatorial chemistry where we're just canvassing these sort of large volumes of data but we're really not progressing very far. We're churning through a lot of data but we're not drilling down deeply enough on any one point to really make a truly substantial discovery.

It is noteworthy that his account, particularly in its emphasis on "churning" and "canvassing," parallels that of his traditionalist peer, quoted above, who described the scientific approach of her commercialist colleagues as "completely mechanical." The similarity of these views implies that, independent of whether one embraces or rejects commercialization, there is agreement that commercialization can work to the detriment of advancing knowledge.

How scientists organize work within their research group depends on their goals. Since status is an important goal for both commercialist and traditionalist scientists, I turn next to the bases on which status is determined in both groups.

Bases of Status

Science, like law and medicine, is a high-status profession, and entrants to such professions are presumably motivated by the desire for elevated professional standing. Therefore their selection of specialties and task domains is likely influenced by the desire to maximize their intraprofessional prestige or the level of honor accorded to a practitioner by his or her peers. In science, the highest status has historically been connected to fundamental discoveries. Between 1901 and 1972, for example, only 14 percent of Nobel laureates received the prize for an invention, while the remainder were for fundamental discoveries.[18] It is therefore no surprise that sociologists of science have long argued that original contributions to knowledge are the chief basis of status in science. Science has in this sense been consistent with Abbott's purity thesis, which holds that professions are organized around a core of abstract knowledge, and consequently, the most esteemed pursuits are "professionally pure," or free of nonprofessional considerations.[19]

This argument works well when applied to historical contexts of science, where economic and power considerations were less influential than professional norms and values, and thus original contributions to fundamental research provided the only means by which a scientist could attain elevated status within the community. Today, however, with the presence of both commercialism and traditionalism in science, the bases of status in science are no longer uniformly aligned with the professional purity of one's work. To fully evaluate this development, let us consider two objective properties of professional status: income and client type.

Both status based on income and status based on client type can reflect either intra- or extraprofessional status. Intraprofessional status is assigned by professionals themselves, whereas extraprofessional status is assigned by the public.[20] Both forms of status merit consideration specifically because, even though status among traditionalists is almost entirely intraprofessional, commercialization entails both intraprofessional and extraprofessional commitments, interests, and social networks.

Income matters because it is tied to occupational position, a basic form of scientific recognition, albeit one traditionally viewed as less important to scientists than awards or visibility. The average salaries of the scientists I interviewed allow us to draw tentative conclusions about the differential bases of status between commercialists and traditionalists and the influence of commercialization on organizational reward systems (figure 1). On average, about half the scientists in the study earn between $130,000 and $175,000 through their nine-month salary. One-third of the scientists in the study earn below this amount. Comparing the distribution of salaries between commercialists and traditionalists, one observes a strong tendency for commercialists to earn more. Approximately 80 percent of the commercialists in the study earn more than $130,000, whereas only 50 percent of traditionalists fall into this category. Twenty-six percent of the commercialists have earnings that exceed $175,000 per year, whereas about 8 percent of the traditionalists fall within this range. Scientists at private universities earn more than their peers at public universities, although this does not explain the difference between commercialists and traditionalists because employment at

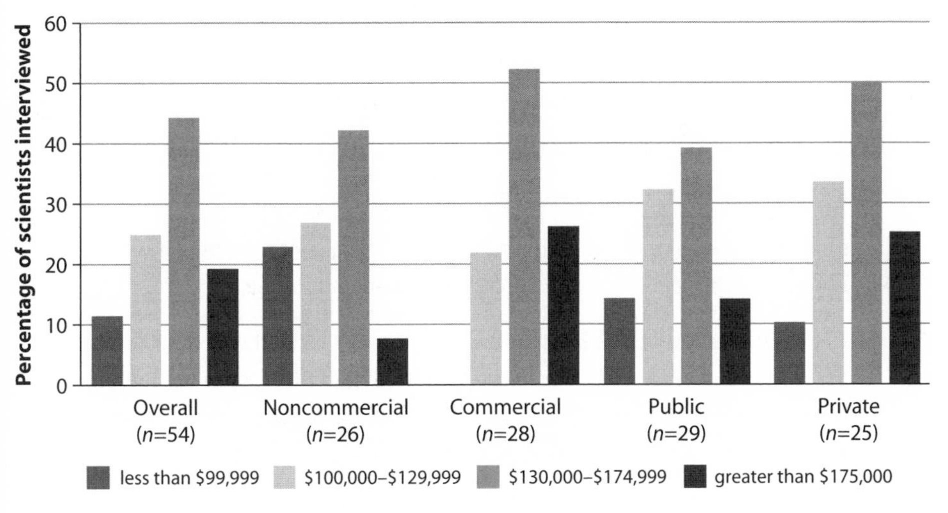

Figure 1. Average Salaries of Scientists. *Note:* Seven scientists did not provide salary data.

private versus public universities is evenly distributed across each comparative dimension of the study.

Income is commonly associated with status outside science, but the data suggest that salary differences provide a basis for intraprofessional distinction among scientists in the same field. Size of income largely reflects what is valued by the administrative strata within universities. Furthermore, although status relations within a profession are rarely established with the knowledge of the income of one's peers, commercialists nevertheless establish their reputations based, in part, on their generation of personal, departmental, and organizational wealth.

To understand how income generation is playing a new role in the establishment of status in science, let us consider the story of an emeritus professor referenced by a number of chemists at one of the universities in the study. By the 1980s, this professor, whom I will call "Hailey," had published nearly ninety research articles and was known primarily for work he had done in earlier decades that described an important set of molecules, although he had not been recognized with any awards or honors at that point in his career. This soon changed as Hailey, in concert with the changes occurring in academic science that permitted universities to patent discoveries, ventured into a new area of research. That work resulted in a discovery and patent that transformed healthcare technology and Hailey's reputation, along with his financial situation and that of his department and his university. Hailey's technology, licensed to a major medical technology corporation, has been used by doctors in several million patient visits and has consistently resulted in annual sales in the hundreds of millions of dollars. It has provided more in revenue to his university and department than most other patents in their history. As one of Hailey's colleagues told me,[21] "There is some disagreement on how appropriate [commercialization] is, but this department in particular has had a patent that probably brought in $10 million to the department from Hailey, and so we would not be in the shape we're in now if we hadn't had that." Another chemist from Hailey's department told me,[22]

> He was always sort of an outcast in the department. I mean there were a few of the inorganic chemists who were close to him and they really loved the guy, but it was only when he started making money for the department that all of a sudden it was . . . "Hailey, Hailey, you're wonderful."

Only after this shift in Hailey's intraprofessional status did he begin to accumulate honorific awards, such as membership in elite scientific societies.

A similar example is found among biologists I interviewed at a public university. One is a professor of biochemistry who,[23] as a traditionalist in a commercially intensive department, feels eclipsed by his commercialist peers, despite his own achievements. Having closely observed the careers of his commercialist colleagues, he says:

> The most prominent guy currently on the campus is ["Walters."] He's the one who gave [this university] the monopoly on [an organism] . . . He's perfectly happy to let [the university] take these damn [organisms] and do what they want with them, but they won't leave him alone, he's worth too much. He started out in complete obscurity—he wasn't even a professor, he was in some adjunct status—and when he discovered these [organisms], the university instantly promoted him and fixed him up. This new building over here [pointing out the window] is his building.

In the stories of Hailey and Walters, we see that income constitutes a basis of esteem, prestige, and respect for commercialists—even when income may not be their motivation. For some, income level becomes the chief basis of their reputation, elevating them from obscurity to prominence within their community.

Even when scientists lack evidence about how much their peers make in salary or royalties, they make assumptions based on news of commercial accomplishments, observable conditions of work, and, in some cases, the ways the commercialists enact their wealth. Traveling through the offices of thirty-two extremely successful commercialists, I found their presentations of wealth quite varied. The most highly rewarded commercialist I interviewed, a biologist whose salary, corporate income, and consulting fees during the past year totaled approximately \$500,000,[24] met me in one of the smaller, more austere offices of any scientist I interviewed, and he noted at one point during the interview that he tries to downplay his wealth. At the other extreme, I encountered a handful of commercialists whose offices were characterized by fine furniture, large wall-mounted flat screen monitors, and in one instance, an elegant (and costly) Italian espresso machine. More publicly, universities and local media often promote the commercial success of scientists, affirming the economic dimension of their social status.

Client type is another key dimension of professional status in which commercialist and traditionalist worlds diverge. We typically do not think of academic scientists as having clients, primarily because the products of their work have traditionally been directed toward other scientists and the scientific knowledge base.[25] But as we consider board representation, officer positions, and consulting work performed by commercialists, we see that professional rebellion involves a new form of status construction in which commercialists value service to socially powerful clients. Because status connotes a system of relations among people, the professional networks of scientists are important to consider, as they reflect differences in the groups through which traditionalists and commercialists seek to exercise their power, alternative bases from which they seek esteem or respect, and thus different bases through which professional reputation may be constructed.

Information I collected on the commercial scientists' corporate positions makes it unambiguously clear that corporations are both regular clients and venues through which commercialists wield considerable power and influence. Overall, the average commercialist in this study has engaged with about fourteen companies for varying lengths of time. In terms of ongoing positions, such as officer or board positions on directorates or scientific advisory boards that involve sustained interaction, the average commercialist has engaged over time in regular contact with seven corporations, some of which they helped to found. Over the course of their careers, approximately one-third of the commercialists have held chief officer positions, two-thirds have served on corporate boards, and nearly all have been on scientific advisory boards. Consulting, although not exclusively a sustained interaction with a corporation, constitutes another frequent nonacademic commitment among the commercialists. As the commercialists' vitae indicate, the companies they engage with range from startups to well-known corporations with extensive resources, such as Becton Dickinson, British Petroleum, DuPont, Eli Lilly, General Electric, Intel, Merck, Monsanto, and Pfizer.

For traditionalists, by contrast, engagement with industry is extremely limited.

Only one has held a position on a scientific advisory board. Only six, or about one-fifth of them, have consulted. Of these, four scientists consulted for one corporation, and the remaining two consulted for three and four corporations.

In terms of intraprofessional status, then, the key difference between commercialism and traditionalism is that commercialists possess broader and potentially more powerful external client networks through which influence can be exercised and reputations constructed. Measures of income and client type indicate that commercialists enjoy a status among nonprofessional audiences that traditionalists do not. In particular, traditionalists rarely consult. As I show in chapter 3, the lack of such connections among traditionalists is a reflection of their commitment to fundamental research rather than lack of opportunity.

The higher salaries and connections to powerful corporations among commercialists suggest that external definitions—those made by nonscientists, especially in commerce—of what is valuable or esteemed among scientists or scientific research differ from what scientists themselves have traditionally constructed as the key basis of esteem. To the extent that the higher salaries are also indicative of what universities regard as esteemed or important, these findings point to a shift in the basis of scientific status at elite institutions. The key question we must now consider is what scientists themselves construct as esteemed.

Commercialists: Delivering Science to Society

The hallmark of esteem among commercialists is societal impact. They believe that the most important thing a scientist can do is to extend the reach of his or her work beyond the scientific community so that it benefits society. Thus, even though most commercialists I interviewed stressed the importance of contributions to basic knowledge, their views of the most esteemed science cohered around notions of impact, utility, application, and extrinsic value. Consider, for example, what a professor of biology I interviewed said regarding the pursuits in science he most esteems:[26]

> To translate our technology and science into society. To make our understanding and control of materials a benefit to people . . . To promote the development of the technology in the public or private sector outside the university, and to expand jobs and benefits of that technology in terms of society and people, citizens of the country and the world in that regard.

I was surprised by this response because he had received the Nobel Prize. Given the contrast between that recognition, which he had received for basic research, and the value he was placing on societal impact, I followed up by asking him to comment on the difference between the impact that commercialists and traditionalists make. He responded:

> It is a different one. We remember Edison, Tesla, and a number of other people who were really known for their inventions. We remember Einstein, Newton, and Darwin for their contributions as well. We don't remember a lot of business people. It just doesn't excel. We're not going to remember a host of people who have run GE. Jack Welch [former CEO of GE] will disappear from our vocabulary in ten years.

Commercialists have constructed a new status system in science in which societal impact—or the control of societal uncertainties through the invention of technological products—forms a basis for scientific immortality. This reconstruction of status drives the overall mode of professional rebellion we observe. "Delivering" science to society as a new basis for status offers a view of scientific work in which extrinsic benefits of knowledge are extracted to solve societal problems and reciprocate societal support for science.

The presentation of this view by interviewees regularly contained an overture to the importance of basic research because it provides the initial basis for commercialization and because it is recognized as the unique claim of university scientists. Commercialists, however, interpret the organizational framework of fundamental research as a niche within which they are able to pursue problems that are neither purely academic nor purely commercial. As an associate professor of biology explained to me:[27]

> A university is the best environment for engaging in pursuits that are abstract, which perhaps do not have an immediate application. But at the same time, if one can keep an application in mind while pursuing these questions, that's better.

Rebellion thus entails, alongside a reconstruction of status, a redefinition of the university as a niche or liminal space between the academic and the commercial, rather than as a purely academic environment.

This interpretation of the university environment is the essence of the commercialist's modus operandi: academic scientists, being unhampered by market-oriented and temporal constraints while possessing fundamental knowledge rarely found in commercial organizations, are positioned to provide socially and economically valuable solutions to societal problems. The moral foundation of this new mandate is one of control. The technological products created by the commercialists I interviewed—such as designing a new mechanism for delivering vaccines, storing information on a device, or treating chronic pain—were designed to address issues related to health, the environment, and general human welfare, in other words, to control uncertainties presented by societal problems. The accomplishment of such a feat is what prompted the Nobel Prize–winning biologist to place the inventor Nikola Tesla alongside the basic researcher Charles Darwin in a pantheon of the immortal. Consequently, for commercialists, societal impact competes with scientific impact as the basis of status.

The majority of the commercialists I interviewed demonstrated this symbolic currency of commercial achievement by juxtaposing their scientific achievements against their commercial successes to argue that commercialization offers greater returns. I spoke with a professor of chemistry whom I will call "Charles."[28] Before our interview I found him drinking champagne with a group of researchers celebrating a recent achievement. He characterized himself as frank, noting this characteristic as a "blessing and a curse." Like other commercialists, Charles juxtaposed scientific and commercial achievement to demonstrate his view that societal impact is the ultimate achievement:

> What I'm most interested in is impact—having an impact on the world. That's the first priority. And then there's a second priority of entertainment, doing things that I find interesting. In graduate school I had been doing [biophysical research]. That work affected a very small number of people that might care about the stuff I was doing, but it turned out the work I did [as a postdoctoral researcher] to automate [a critical scientific technique] ended up changing the course of the world in a significant way, and, in fact, in billions of people. So it's just not a tough call. If you've got so many hours in the day to spend, how do you want to spend them? And if you can do something that has a big impact, then for me that time is better spent than if you have something that has little or no impact.

"There are only so many hours in the day" was a familiar refrain as I interviewed commercialists around the country. Contributions to fundamental research no longer wield the exclusive claim to eminence in science. The bases of status in science now include activities and outcomes ancillary to academe's distinctive claim of fundamental research, such as the development of commercial products, solutions to nonscientific problems, and economic development and the creation of jobs. Consequently, today commercial activities compete with or replace original contributions to knowledge as the basis of status in the moral order of commercialist science, a reconstruction that drives professional rebellion.

Traditionalists: The Luxury of Being Paid to Think

As we have seen, and as existing research confirms, traditionalists view original contributions to knowledge as the most esteemed scientific pursuit.[29] For them, status is derived from transforming the way one's peers think about their field, and they place a premium on originality and priority in discovery. Therefore, the only means by which a scientist can enhance his or her status within the scientific community is by making a contribution to science that provides other scientists with new tools and new knowledge. As a professor of chemistry explained when I asked him what pursuits in science he most esteems:[30]

> Creating new truth. We have the luxury of being paid to think and the most valuable product is new ways of understanding how things work . . . I have never or almost never tried doing anything except curiosity-driven experiments.

The preceding comment embodies the purity thesis in its emphasis on the curiosity-driven discovery of new knowledge, and also because it hints at derogating tasks or considerations that constrain the focus on "pure" research. An assistant professor of biology invoked the purity thesis more explicitly when I asked her to explain what she values most in science:[31]

> Discovery-based work is personally most exciting to me. It's also what is not duplicated anywhere else. If it's very clearly goal-oriented—likely to result in something either profitable or medically useful in the short term—then I feel like that kind of work is probably naturally pursued vigorously by the private sector . . . I have a sense about science that is explicitly short-term-goal oriented—like producing a device, technology, or a drug—that it is always trying to find the fastest route to something of use, as opposed to seeing what the most interesting thing is, that's true.

Her comment underscores a concern shared by traditionalists: that commercially oriented science sacrifices the core professional standard of truth. Thus, in contrast to commercialists, whose rebellion involves a redefinition of the university, traditionalists espouse the pursuit of truth, free of nonprofessional interests, as the mandate of the university. When I asked an assistant professor of chemistry what he most esteems, he said:[32]

> I would have to say it's basic research to the extent that a lot of science and technical research can be done just as easily in industry. I think that the roles should be pretty separate from one another, or ideally, could be separated from

> one another. So the sorts of things that industry won't do because they're not profitable but are nevertheless interesting for fundamental reasons, I think, are really what the mission of a research university ought to be.

A noteworthy feature of these accounts is the extent to which traditionalists' constructions of esteem are framed in opposition to characteristics of commercialist work. Traditionalists stress that traditional science involves openness and contributions to the scientific community, whereas commercialist work entails elements of secrecy and commitment to societal rather than scientific impact. They view profitability as a potential threat to science, whereas commercialists view it as a criterion by which scientific work can be judged.

Three preliminary conclusions can thus be drawn based on the ways the commercialists and traditionalists I interviewed constructed status in science. First, their expressed views suggest a contest between moral orders, particularly to the extent that traditionalists view the embrace of commercial activities as a threat to their view of how science should be done and their ability to pursue fundamental research. Traditionalists thus exhibit a threatened social-psychological attitude, whereas the attitude of commercialists is unthreatened; they view their position as enabling a preferred, entrepreneurial course of activity. Second, the views indicate that commercialization is not as widely embraced by scientists as some researchers have argued. The traditionalists' accounts varyingly reflect tolerance, reluctant acceptance, or rejection of commercialization, but rarely wholehearted acceptance. Third, these accounts reveal a competition between reward systems wherein one subset of scientists seeks status through contributions to the scientific community while the other pursues status through contributions that attempt to control societal uncertainties and bring indirect, nonscientific returns to the scientific community.

Visibility

Professionals' perceptions of who benefits from their work highlight, in a general sociological sense, the ways reputation within a profession is constructed. In the context of science, these benefits—often framed colloquially by scientists as "impact"—are conceptually tied to visibility, or the extent to which a scientist and his or her work are known within the scientific community.

Visibility is a critical component of a scientific career, for rewards are predicated on contributions, and scientists cannot be rewarded if the relevance of their discoveries to scientific advance never become known. How scientists communicate their ideas is therefore critical to both the operation of the reward system and individual careers. Visibility and its antonym, invisibility, connote a continuum representing the spheres of influence potentially attainable over the course of a career. At one end are invisible scientists whose contributions are unpublished, uncited, or constrained to functions ancillary to research, such as training and teaching. Such scientists have limited and exclusively local influence. At the opposite extreme one finds the scientific immortals—scientists whose transformative contributions to knowledge create visibility and influence that lives beyond them through eponymous laws, theorems, or branches of science. More common occupants of this extreme of visibility are scientists who have received the Nobel Prize or are members of the National Academy of Sciences. As sociologists of science find, most other awards are not highly visible to the national community of scientists.[33] For neither the invisible nor the extremely visible scientists, other accomplishments, such as citation and appointments at elite universities, provide traditional means by which visibility may be enhanced.

The visibility continuum is representative of the stratification system in science. As sociologist of science Harriet Zuckerman notes, "A small number of scientists contribute disproportionately to the advancement of science and receive a disproportionately large share of rewards and resources needed for research."[34] The majority of the population of scientists is thus clustered toward invisibility.

Visibility, it should be noted, has a temporal aspect. One's position along the continuum varies over the course of a career, as the acquisition and maintenance of a reputation requires making ongoing contributions. What is more, visibility may expand or decrease based on advantages experienced early in the career. The principle of the "Matthew Effect" suggests that those who achieve, or are associated with, eminence early in their careers will accrue disproportionately high visibility over the course of their career because visibility can be converted into resources for research.[35] Scientists of lesser repute are unable to accrue visibility in this fashion. We may thus conclude that visibility is a dynamic aspect of a scientific career, and that although scientists may attain high visibility, sustaining and exceeding that reputation is difficult after one's "climb to the top" plateaus. In asking scientists who benefits from their work, we access their views of where they have had the most influence. In doing so, baseline conventions concerning what is attainable may be identified in each moral order, indicating how each group responds to an environment in which progress and development, not stability, is the normal expectation.

Commercialists: Mundane Accolades

For most scientists, publishing an article in *Nature* or *Science* remains a dream, not easily achieved. It is one means by which a scientist can achieve remarkable visibility, separating himself or herself from the majority of the community—except when one is surrounded by departmental peers whose expectations for proving one's worth require publishing in top journals. How does one elevate one's reputation in an environment in which an otherwise extraordinary accomplishment seems ordinary or mundane? In the case of commercialists, they rebel, staking out other forms of visibility to set themselves apart from their peers.

Only one-third of the commercial scientists I interviewed referenced visibility within the scientific community in their comments. Among these scientists, only two exclusively noted the scientific community as the group for whom their work has had the largest or broadest outcome. This relatively low number of commercialists who emphasized scientific impact is particularly surprising, given that nearly all of the commercial scientists interviewed are highly productive and highly cited researchers at well-recognized American universities. Overall, the commercialists' explanations of their influence were predominately framed in terms of societal benefits. That is, their comments on visibility contained no referent to any individual, group, or body in the academic scientific community. Representative of this emphasis is the response of a professor of chemistry when I asked who benefits from his work as an academic scientist:[36]

> You have to realize that, I would say, in a twenty-year period, I have been involved with probably something between ten and fifteen startup companies, and the fact is that many thousands of people have jobs right now because of my technological ideas and the efforts that arose from those. And that's still going on right now. Okay? Because I'm typically working with five companies at each point in time.

It is notable that he did *not* mention the four hundred articles he had published. Nor did the majority of other commercialists emphasize this form of visibility in the scientific community. Commercialists, especially those who received their PhD before 1980, have commercialized work in ways that have large-scale and even global reach. Considered alone, it is not difficult to understand why scientists who have more than three decades of experience in academe look back and cite commercial, rather than exclusively scientific, accomplishments as evidence of influence and impact on the world, because these scientists underwent professional socialization in an era in which their role models were unable to claim such accolades. In this respect, construction of visibility among commercialists entails an element of calculability, in that the quantifiable value of one's work is materially expressed through the universality of one's inventions and economic indicators of visibility such as jobs created and licensing revenue.

However, commercialists' emphasis on their commercial achievements to the exclusion of their scientific impact is difficult to reconcile, given the level of success suggested by their appointment at premier research universities. One explanation of this pattern is that the majority of scientific contributions, even among the most successful scientists, are viewed as incremental or as having only a minimal additive (rather than transformative) effect on knowledge. According to this interpretation, achievement in an intensely competitive environment is viewed as routine even though such accomplishments set these scientists apart from the majority of academic scientists in the world. This is seen in the way commercialist scientists so frequently discussed such accomplishments in relatively mundane terms. A professor of electrical engineering,[37] for example, stated,

> I've certainly done in my own career the usual kinds of things, published hundreds of papers and gotten the usual kinds of research awards and gotten lots of teaching awards and things like that.

Similarly, Charles explained that

> you check off the boxes of the things you've accomplished—like so many papers, so many federal grants, talks, blah, blah, blah—then one of the boxes is "started a company," and that adds cachet to your resume rather than a bad thing.[38]

Or, in the words of an assistant professor of chemistry who was up for tenure at the time of our conversation,[39] "Publishing in top journals is great, but it's pretty common. It doesn't feel like a really unique thing."

Having already achieved heightened visibility through publication, citation, and the achievement of tenure within elite scientific departments, commercialists redefine originality such that the scientific pursuits through which visibility is initially acquired are attributed as normal and incremental, whereas societal impact is framed as unique and providing a new form of visibility. They rebel by substituting new modes of visibility and meanings of influence that downgrade traditional means of establishing a reputation. Commercialization provides a new outlet for visibility and rewards once one has "arrived" and solidified one's place in the elite strata of science, particularly because the difficulty of acquiring more visibility only increases with the establishment of one's position in this environment.

The prestige commercialists assign to external visibility is so potent that it forms the basis of aspiration among commercialists who are in earlier stages of their career. The commercialists I spoke with who had received their PhDs after 1980, like their more senior commercial peers, emphasize societal impact, but their perception of visibility is framed in the future tense. A comment made by Rose, a professor of chemical engineering whose views we encountered in chapter 1,[40] illustrates this point:

> There are multiple constituencies that benefit from my work. We're actually developing material systems which we hope will ultimately make some difference in providing a pharmaceutical advantage for the delivery of drugs or in providing material systems that in some way provide medical development that can help others . . . We're also working on energy problems . . . These are the kinds of things that we're looking at, and in that case, I think, there's benefit to the greater world in terms of addressing energy issues and energy problems.

Because Rose and other younger commercialists stress societal benefits in terms of the future, their comments are frequently characterized by multiple possibilities rather than concrete results. This pattern is particularly striking, given the accomplishments they have already achieved, which they could also choose to mention while discussing visibility. Each of these commercialists has achieved tenure in an elite department, indicating that in the eyes of their peers their contributions to their field have been influential. The desire to stand out as unique among the elite leads these scientists to make sense of their present status in light of a future vision, as the perceived rewards obscure or outweigh their perceptions of past accomplishments.

Traditionalists: Bricks in the Edifice of Science

A metaphor for scientific knowledge scientists occasionally employ is an edifice of truth, built brick by brick. It is upon foundations of knowledge, laid over time by the community, some more secure than others, that new bricks of knowledge are added in a cumulative fashion to fortify structures of knowledge. As philosopher of science Thomas Kuhn noted, at times the edifice changes through revolution, when new ideas destabilize existing foundations of knowledge and require the development of new wings of truth. Traditionalists would rather be known as demolition men and women than as bricklayers, but few scientists can lay claim to such a title. They discuss their influence and reputation in varied ways, but the predominant mode of visibility emphasized is reputation within the scientific community.

We begin with "Lloyd,"[41] a professor of chemistry in his sixties whose curriculum vitae offers extensive objective evidence of visibility. Lloyd has published 285 peer-reviewed articles, which have collectively received almost 9,000 citations. Seated in a large office with stacks of papers on numerous tables, Lloyd told me:

> I've been in this business for thirty-six years. I've produced a lot of graduate students. I've been recognized by my peers with a lot of prizes. I have reshaped the way people think about a lot of classic problems, and I think that I am one of the most important people in the world in my field. People use my terminology, they use my books. I've enjoyed doing it, and I'm quite proud of the work I've done. So, I'm not shy about this.
>
> *Who do you think benefits from what you do?*
>
> I don't care. But, I know that . . . a lot of the [scientific] techniques that I've developed are used in practical areas . . . I think that a lot of people that I don't even know about use what I do, and I'm not that concerned about the applications.
>
> *Why don't you care who benefits from your work, as you say?*
>
> Because I know it's good. I know it's fundamental. And I publicize it, but I don't go looking for anybody other than grant agencies to pay for it.

Visibility for traditionalists like Lloyd is solely predicated on the extent to which their research contributions exert observable influence on the scientific community. Thus citation of one's work—acknowledgment of adding important "bricks" to the edifice—and honors conferred by the community provide the most important metrics of success for a traditionalist. Lloyd's account further reveals how traditionalists actively seek to enhance their reputation through self-promotion—in contrast to the methods pursued by commercialists discussed earlier. Traditionalists seek to increase their visibility only within the context of the scientific community and reject alternative bases from which reputation could potentially be constructed. Professional purity means being known exclusively for influencing scientists, not the market, firms, or societal problems.

Visibility derived from commercial influence plays no motivating role in traditionalist careers. This is particularly evident in the case of a molecular biologist who,[42] more than any other traditionalist in the study, claimed a substantial societal impact resulting from his work. Yet despite the magnitude of the product, for him influence beyond the scientific community remained only a secondary metric of satisfaction.

> I think society benefits tremendously. In my case, my research has been purely motivated by an interest in basic knowledge, and yet very early on, with the discoveries that I made, it had direct applications to biotech . . . One-third of the world's supply of [a disease treatment] is now made . . . following the basic pathway . . . that we worked on in my lab. I never would have guessed that.
> *Would you say that's the biggest impact you've had as a scientist?*
> No. Commercially, yes. But intellectually, no. Intellectually, to me the most satisfying thing is the basic knowledge that we've gained. That it had a practical application is gravy, but it certainly isn't the core of what I find most appealing.

Such a construction of influence is the inverse of how commercialists perceive their work. Although he did not take a direct role in the commercialization of his discovery, the biologist is able to claim a societal impact of equal magnitude as the technological and economic impacts of his commercialist peers, yet he frames this as peripheral to how he thinks about his influence.

Many of the traditionalists I interviewed are known for transformational discoveries, evidenced by their achieving not only tenure in elite science departments but also citations and honors, such as membership in the National Academy of Sciences. But high-impact discoveries are episodic rather than routine events in science. Consequently, traditionalists emphasize graduate training as an important sphere of influence. Transformational discovery may be the hallmark of traditionalist visibility, but influence on science through one's graduate students is the constant of a career. Consider my interview with a professor of chemistry,[43] a leading scientist in his field whose visibility is indicated by his membership in the National Academy of Sciences. When I asked him who has benefited from his work, he listed his peers and graduate students and then said:

> There are only a few of us who are going to do something that in Kuhn's language shifts the paradigm. Most of us are going to put little bricks in the edifice, and we do a piece of research and it's useful and it advances science, but there's this huge multiplier effect from our graduate students. I have graduate students who are teaching at research universities, at liberal arts colleges, government labs, and in industry, and each one of those is teaching a different cohort of people and in some cases doing things that are profoundly important. So if you were trying to step back and ask from society's point of view, "Why do you want to have research universities?" it's to discover new knowledge, but it's to transmit that knowledge into society at large and in particular get the next generation up to speed so that they can go engage in that enterprise.

For traditionalists, therefore, while individual visibility is variable over the course of a career, influence, including societal influence, is always viewed as collective. Societal impact is not a motivating factor for traditionalists so much as it is viewed as the product of collective contributions. Traditionalists frame societal influence collectively and indirectly through reference to the profession at large and their students. Although some traditionalists are able to cite specific instances of aspects of their work being incorporated into practical or industrial applications, most believe that the grand problems society faces will only be solved through the collective influence of the scientific community. As one professor of chemistry responded when I pressed him to reflect upon the societal aspects of his work:[44]

> My work is not focused on the societal impacts of our research; it is focused on understanding the science, with the faith that this will ultimately translate into bigger consequences that may be read by others—other chemists or engineers or doctors or researchers in other areas that need a substance that has certain properties that it so happens that we know how to make . . . [Solving societal problems] happens through the accumulated wisdom of many different kinds of scientists. What gets you up in the morning is, I think, not these things that might result from our work; it's cracking the little nuts that are on our table at the moment. That's the way that I look at it.

Dirty Work Designations

The pursuits that scientists shun or consider trivial indicate conventions of moral orders that designate negative role sets, or tasks that depart from desired modes of conduct. Reward systems are reflected in such conventions, in that the tasks scientists designate in this manner bear qualities considered intrinsically unrewarding or extrinsically unworthy of professional honor. Traditionalism and commercialism share the designation of three types of activity as dirty work: administrative responsibilities; "job shop," or directed scientific work; and incremental science. No distinctive differences were found in the level of emphasis scientists gave to any of these activities, nor were any differences found in the scientists' rationalizations of the triviality attributed to these three categories.

Administrative work comprises activities, ancillary to research, that faculty deem either boring or "necessary evils." University committee work and the administrative aspects of running a laboratory, for example, were referenced by scientists as "pure baloney" or the "crap quotient you just have to do." These tasks constitute what sociologists of work would refer to as the "dirty work" aspects of science, or practices that are legitimate aspects of the academic role that are regarded as tedious or unrewarding but are nevertheless required.[45] As an associate professor of biology in her forties told me,[46] "I don't think committee work is not valuable, but it takes a tremendous amount of time so it's a drain. One quarter of [science] is painful, bureaucratic nonsense." The performance of such tasks carries no threat of sanction from one's peers because it is recognized by all as a necessary, albeit unrewarded, component of one's academic duty.

In contrast to trivial but otherwise normal aspects of academic work are tasks that scientists believe should be shunned. These are activities that may be permissible but are regarded as inconsistent with or not in the best interests of the advance of knowledge, training, or teaching. The activity representative of this category is what scientists call "job shop" science, which denotes research on a defined problem or performance of a specified task at the request of an organization such as a firm, corporation, or research agency. The issue is not the nature of the external organization but the way the work is organized. Thus, for the commercialist, job shop work may be rejected for lack of interest or because it does not advance the work being done by one's research group. As a commercialist chemist explained:[47] "I won't do anything which is just work for hire. Somebody calls me up and says they want me to examine a patent and testify? I say no. Because I'm not interested in it. If I'm not intrinsically interested in it, I don't do it." The traditionalist may reject job shop work for similar reasons. As Lloyd, traditionalist chemist we heard from who is "not shy about" his achievements, explained:[48]

> I would never work on somebody else's problem. If there is a request for proposals in a very carefully defined area . . . if they're too directed I don't want to do it because someone else thought they could decide what was important. I'm not willing to trust someone else's judgment . . . I have the luxury of working with the best graduate students. I'm not going to subject them to some narrowing experience.

Scientists shun activities they view as threats to their autonomy or authority, or as potentially degrading the training of graduate students. In contrast to the localized nature of attributions of dirty work, shunned tasks are more frequently subject to individual discretion. Scientists may not gain the universal respect of their peers for consulting or working on targeted research, but neither would they be subjected to informal or formal sanctioning on the basis of their peers' personal preferences.

The most prominent dirty work designation among commercialists and traditionalists is incremental science, or the publication of research contributions that have little impact on knowledge. The irony is that this form of dirty work is a routine feature of science. Scientists I interviewed regard this type of practice as normal, acceptable, and not even wrong, but nevertheless beneath the caliber of science expected of them. An assistant professor of biology in her first year at an elite private university captured the emphasis both commercialists and traditionalists place on incremental research when she said:[49]

> I think the power of modern molecular biology techniques is such that it's really trivially easy to turn the crank and publish something that is simply not illuminating. It's not that it's wrong. It's not that the data are even of poor quality. But it's like writing the alphabet over and over again. The problem is that it's good for students to publish papers, and it's good for labs to publish papers.

Scientists' views of incremental publication thus represent an expression of the ideal conduct of a scientist as it exists in the institutions studied. The scientists are encouraged to conduct high-risk, high-impact science, but they must nevertheless publish, and in the best journals possible. Otherwise, in terms of the outcomes of their research, there is little that sets their visibility apart from that of their peers with whom they compete.

What ties administrative responsibilities, "job shop" research, and incremental publication together, apart from the fact that none has any impact on visibility, is that scientists almost inevitably end up engaging in these pursuits, despite preferring to avoid them. Dirty work designations do diverge between commercialists and traditionalists, however, with regard to two patterns of conduct: book writing and commercialization.

Commercialists: Overvalued Academic Exercises

The key dirty work designation that, for commercialists, differentiates them from traditionalists is their views on writing books. From the perspective of the commercialist, this activity does little to advance knowledge, is influenced by a concern for profits, and is a poor investment of one's time. All three points emerged when I spoke with a professor of chemistry who characterized both textbooks and monographs as "not an effective use of time."[50] When I asked him about activities he shuns, he replied:

> Writing books would be one activity I shun. There are some people who feel that writing books is a useful pastime. I feel that is something that is not worth

> the time invested in terms of the impact. Most books now are done by publishers who are more out to make a quick profit. So a lot of them are "short time" books, like hot topics, but, you know, a lot of time invested and yet, you know, they're basically out of date in a short period of time.

The commercialists I spoke with varied in which form of book they rejected, but their critiques of book writing as an activity were similar across interviews. In talking with commercialists, I was struck by the extent to which their critiques of books overlapped with some scientist's critiques of the commercialization of research. A commercialist professor of biochemistry responded when I asked him about this overlap:[51]

> I see a lot of wasted effort on producing yet another low-level textbook that might or might not outsell the others. I do not think of that as a scholarly pursuit. I view that as largely a selfish endeavor.
> *Could one make a similar critique of the commercialization of research?*
> Well, I'm not going to argue with that, actually. I'm going to say that when I participate in [my company], I don't view that at all as being part of my job as a professor. I view that as something that I'm doing with my free time, if you will, off salary, different, even though the university has a stake in the success, because they own the patents . . . [But] let me finish with the books, because the books, to me, are overvalued as academic exercises. I think that writing yet another textbook contributes virtually nothing to academics and the progress of humanity and therefore cannot be viewed as a fitting enterprise for an academic scientist.

That commercialization and book publication share similar qualities is not lost on commercialists. In chapter 4, which examines how commercialists legitimate their role, we see that the time investment and potential for profits associated with book publication lead many scientists to compare their commercial activities with writing books. Commercialists construct book publication as a dirty work task because they perceive the activity as lacking societal impact and offering only minimal returns to science at the cost of commitment.

Traditionalists: A "Pact with the Devil"

While commercialists regard book writing as dirty work, traditionalists construct commercialization in a similar light. They tacitly accept commercialization by virtue of its overwhelming presence within their departments and their institutions. Commercialization occupies an unmistakable presence on the campuses of the institutions in this study, and departmental funding and building construction serve as material reminders of the presence of commercial culture. As a professor biochemistry noted:[52]

> When you walked into this building, you walked by a framed patent on the wall. It's actually ["Watts's"] patent, and that is there to tell you, you just walked into "the House of R&D"—this building. There is the constant reminder that the university profits in real ways from this enterprise, but there is something slightly apologetic about that. Everybody realizes that this is a very risky business and that you sometimes have to hold your nose and plow through it.

Acceptance of the presence of commercial culture, however, does not equate to participation or encouragement. In this respect, for traditionalists, commercialization is simply a property of one's environment, something to be put up with, similar to other forms of dirty work. Consider, for example, the words of a professor of chemistry whose view of commercialization reflects the ambivalence characteristic of traditionalists:[53]

> The ability of academics to get patent protection for discoveries made with NIH, NSF, and DOD money was a pact with the devil. I think it will always be the case, as with textbook writing and other things, that it can be overdone. And it is overdone by some people. There's downward pressure on teaching loads because these businesses are taking up peoples' time, and so they say: "How can I do my job properly if I have to teach all these stupid undergraduates?" You get this kind of distancing from what is the central core function of the university as a result of these involvements. Where do you draw the line? I mean there's a lot of money to be made here by a few people, but most of these things never pan out.

Traditionalists are attuned to professional rebellion—in this comment, the "distancing" from the core function of a university—and its consequences.

In their general orientation toward commercialization traditionalists recognize three negatives it causes: the opportunity costs external to science derived from exclusively fundamental academic science, the risks posed to academic science by virtue of engagement in commercialization, and the rarity of instances in which the human and material resources invested in commercialization bear fruit that merits commitment to commercial objectives.

Consequently, traditionalists shun commercialization. For one, they perceive commercialization as an excessive distraction from the key objective to which scientists, in their view, should be committed: advancing knowledge. Nearly all of the traditionalists expressed this, but few were as adamant as an associate professor of chemistry I interviewed.[54] When we met, I noticed both a coffeemaker and a bottle of whiskey in the corner of his office, which, coupled with his disheveled appearance and his passionate demeanor, led me to believe he spends most of his time in his office and lab in a dogged pursuit of science. When I asked him if he sees commercialization as a legitimate activity for academic scientists, he responded:

> It's hard to balance one's commitment to pure science and to commercialization would be my judgment. I think that if one is going to try to commercialize something and spend a long time there, that is going to have an impact on your ability to pursue pure science. I think that academic scientists should be pursuing pure science. Commercialization should be if something happens to fall in your lap, you know, perhaps you should consider commercializing it just for the good of the scientific community, in that it provides society a payback, if you will, but it is a distraction from thinking about pure science, in my view.

The notion that commercialization should be something that "falls in one's lap" was a prevalent theme among the traditionalists I interviewed. For these scientists, commercialization is a practice unworthy of commitment unto itself. It is an activity commercialists should pursue if the opportunity presents itself and the outcomes are beneficial to science, but it is not a legitimate objective in its own right.

Traditionalists also shun commercialization because of conflicts of commitment and conflicts of interest, which from their perspective threaten the integrity of both universities and science. A professor of mechanical engineering,[55] for example, sees commercialization as problematic for one's commitment to departmental responsibilities and one's graduate students. For her, the question is, "How often are you here, and if your outside commitments, whatever they are, take you away from here an unduly amount, then who's minding the farm?" However, most of the traditionalists I spoke with were more concerned about conflicts of interest. A professor of molecular biology expressed this view:[56]

> There ought to be a limit. At some point you want to have people that don't have a commercial conflict of interest in a given statement. If somebody from the outside world comes to the academic and says, "Are genetically modified foods dangerous?" you don't want to only have people that answer that have a commercial interest in genetically modified foods . . . I have no dog in that

> fight. I'm not making any money. I'm not trying to be on the board of Monsanto. If somebody hears my opinion—and I must say all scientists are very reluctant to give an opinion because we all really know we don't know as much as we should know to be giving good advice—but at least I'm not conflicted. If every biologist at [this university] has some commercial connections, at what point can the community as a whole trust their judgment? . . . Maybe this is a hopeless dream. Maybe it's a hopeless dream to have experts that can venture opinions on subjects that they're knowledgeable in but they're not conflicted in. Maybe that's just a naïve, impractical idea now, but I don't think so.

Traditionalists shun commercialization due to its perceived threats to organizational and professional processes and outcomes. Organizationally, commercialization may influence departmental responsibilities such as committee service and teaching loads in ways that place the burden on traditionalists. Professionally, the concern centers on the integrity of scientific knowledge as it is conveyed in journals or to the public, the training of future scientists, and the preservation of pure science.

Conclusion

The moral orders of traditionalism and commercialism are anchored by purity and rebellion, respectively. Purity is evinced by traditionalists' commitment to historically conventional elements of an academic career. Rebellion is evident in commercialists' pursuit of status through new means. These new means involve formally rational substantive values: a mandate to control particular societal problems, the adoption of more efficient, hierarchical modes of organizing the scientific division of labor, and materially calculable notions of what it means to have an impact. These departures result in a contest between these two orders of science, embodied in divergent modes of organizing work and divergent bases of status, constructions of visibility, and designations of dirty work. The study's findings reveal that the purity thesis—an argument proposed by sociologists of science and professions—is unable to account for commercialization as a basis of intraprofessional status among scientists who occupy the ranks of the elite. Abbott states that "the academic professional's high status reflects his exclusively intraprofessional work." In his view, engagement with the "squalid reality" of nonprofessional issues mocks the "pristine abstraction" of professional work, and therefore professional engagement with applied or impure science would be considered professional defiling.[57] Yet as the evidence here clearly indicates, the "dirty work" at the front lines of professional practice that most traditionalist scientists avoid and regard as irrelevant is, in the eyes of the commercialists, the basis of power, eminence, and achievement. In the case of the commercialists, therefore, the purity thesis does not hold.

At this point in our examination of the operation of a commercially oriented reward system in higher education, we have considered the norms that faculty espouse and the status systems associated with such norms, thereby identifying the key elements of professional ideologies of traditionalism and commercialism. This prompts new questions: How do some faculty end up embracing commercialism? Why do some adhere to traditionalism in contexts where commercialization is pervasive? What are the implications of commercialization for professional identity? The next two chapters address these questions. I now turn to the social mechanisms by which scientists embrace and avoid commercial trajectories.

CHAPTER THREE

Embracing and Avoiding Commercial Trajectories

Walk onto the campus of nearly any research university in the United States and you will inevitably encounter signs for entrepreneurial competitions, hackathons, and seminars with titles such as "Turning Technology into Startups" and "Lunch with a Venture Capitalist"—all of which seek to promote innovation and unleash the commercial potential of undergraduates, graduate students, and faculty. Research universities strongly encourage commercial behavior, particularly in science and engineering, yet it remains unclear why some faculty embrace commercialism while others embrace traditionalism. Factors such as prior achievement, institutional prestige, local commercialist peers, and an institutional commercial culture are environmental properties conducive to the adoption of commercial behaviors.[1] What, however, explains why scientists embrace or eschew commercialism when these factors are "held constant"? Given that both commercialists and traditionalists work in university environments characterized by extensive resources for research and commercialization, a complete answer requires that we examine how commercialist scientists make sense of their commercial trajectories and how traditionalists explain why they maintain career paths oriented around traditionalism.[2]

What are the social mechanisms by which scientists embrace and eschew commercialization? Extending sociologist of work Dan Lortie's approach to occupational recruitment, which discusses the process of choosing a career path in terms of attractors and facilitators, I examine commercialism and traditionalism as career paths that have distinctive attractors and facilitators.[3] Attractors are features of a career path that offer advantages or benefits to a scientist, whereas facilitators are attributes of the scientists themselves that impel them in one direction or another. We begin by considering why scientists became commercialists before turning to traditionalists' explanations of why they avoided commercial trajectories.

The Making of a Commercialist

No singular factor explains why scientists end up commercializing their work. Funding constraints represent one of the most common factors emphasized in some scholarship, yet none of the scientists I interviewed cited funding issues as important to their decisions to commercialize their research.[4] Management scholars have certainly attempted to identify the conditions that give rise to a commercial trajectory; while influences such as institutional prestige, technology transfer office staffing, and commercialist collaborators, for example, are indeed important influences on a commercial turn, my interviews with scientists suggest there is more to it than any of these ready explanations would suggest. The scientists gave widely varying explanations of what led to their first patent. For the commercialists I interviewed, three facilitators prompted the pursuit of commercial opportunities: professional socialization, social origins, and social serendipity. Receiving equal emphasis in their accounts were four attractors that lured scientists toward commercial practices: material benefits, ethical boundary work, societal impact, and tangibility.

Professional Socialization

Institutional and individual exposure to commercialization among *all* scientists who entered academe prior to 1980 was low, despite the slowly emerging signs of commercialization in academic science during this era. Only one commercialist who completed his doctorate before 1980 did so under an established commercialist, and the experience resulted in a series of patents during the 1970s that were critical to developments in biotechnology. Yet, several commercialists who trained in this era did not embrace commercial behaviors until well after they had achieved tenure, and they emphasized in interviews that commercialization was "unthinkable," "tainting the well," or seen as "a bastardization of science." As one commercialist explained to me,[5] "In those days, you definitely weren't in it for money. You were in it because you wanted to find out something new." For these scientists, other facilitators and attractors ultimately pulled them into a commercial role.

For scientists who completed graduate school after 1980, however, professional socialization was one of the key mechanisms facilitating a commercial trajectory. During this time, science became more structurally and culturally differentiated, a process driven by accelerated commercialization and new organizational forms, such as university-industry research centers, which fostered tighter links between university scientists and industry. Twelve of the seventeen commercialists I interviewed from cohorts of scientists who trained after 1980 received training from established commercialists as graduate students or postdoctoral researchers. For these scientists, exposure to commercialization during socialization was strongly influential in their own identification with the commercialist role. Commercialization was elemental, or a standard component of the scientific life, and modeled by their advisors. As a professor of biology said of his advisor:[6]

> I saw that he was able to wear these different hats . . . and that was formulative in the sense that it taught me how this could be done and showed me that it actually could indeed be done, and so that was great. Everyone [at the company] knew that I was working with him. He was the big boss, and so I had carte blanche and it was just a wonderful experience. So that was very influential.

Other commercialists I interviewed who characterized their graduate experiences as "intensely fundamental" or "hard-core science" described similar experiences, but with their postdoctoral advisors. Some commercialists started along a traditional path as doctoral students, but no matter what view of commercialization they developed during this period, having a high-status commercialist advisor conferred validity and prestige on the commercial trajectory. Indeed, most of the commercialists I interviewed who trained in the decades following the Bayh-Dole Act emphasized the eminence of their advisors and the prominence or size of those advisors' companies, and the key process they emphasized was learning the new role from an eminent commercialist. For example, when I asked a professor of chemistry what led to his embrace of a commercial career trajectory,[7] he explained:

> My basic philosophy about doing transformative science and the sweet spot of conceptual advances that also lead to practical applications was inspired by my PhD advisor. He won the Nobel Prize in chemistry. He started his own company, and many of his inventions and discoveries are now commercialized. I had a very good role model. It was learning from his example more than any direct conversation about how to do it.

Socialization to commercialization under an established commercialist occurs both indirectly and directly as role aspirants emulate their mentors' standards of achievement. Therefore, even when commercialization is not an explicit component of the content of professional socialization, or is not even discussed, its legitimacy is taken as self-evident by virtue of example. When one is socialized to commercialization by an established commercialist, the acceptability of commercialization is embraced without question. Scientists-in-training learn to view commercial behaviors as normal aspects of a scientific career. The operation of commercial rewards thus influences scientists' aspirations alongside traditional rewards, but absent the unfavorable definitions assigned to commercialization by cohorts of scientists who were trained prior to the onset of commercial culture.

Socialization may have driven most of the more recent cohorts of commercialists that I interviewed toward this role, but not everyone trains under a commercialist. Indeed, earlier cohorts of scientists underwent professional socialization before commercial culture had fully emerged in academe. For these groups, other facilitators—such as social origins—and attractors would lead them down a commercial trajectory.

Social Origins

Socialization in early stages of life also shaped the experiences of some commercialists. Commercialists I spoke with who trained before 1980 frequently laced the accounts of their first commercial acts with overtures to their nonscientific pasts or to having long-standing dispositions toward interest in applied science. For some within this group, the desire to discover and create technological solutions to societal problems was framed as a personal disposition that influenced their selection of research problems. For a minority, it was a rationale for founding companies or patenting research before the Bayh-Dole Act. For these scientists, social origins that celebrate or value practical utility do not cause one to become a commercialist; they are social-psychological mechanisms that ease one's entry into a role. Consider, for example, a physicist in his fifties now working as a professor of electrical engineering.[8] When I asked him why he commercializes his work, he began by stressing a disposition to utility from an early age.

> My experience may be slightly unusual in that regard. As a teenager, I became quite interested in science and electronics. I was fairly good at doing subjects like physics. I was doing a combination of playing about with electronics as a teenager and doing physics in high school, and then when I went to university . . . When I looked around for somewhere to do graduate work . . . some places really turned me off. The guys who were more over into pure physics didn't really appeal to me, because I couldn't figure out *why* they were doing what they were doing . . . They didn't seem to me to have any connection to stuff that was particularly useful.

Whereas some commercialists I interviewed spoke of an early interest in the application of science as a factor motivating their career decisions all along, others made sense of their present commercial activities in light of their pasts. They suggested that their commercial activities were derived from context-specific values they were socialized to in earlier stages of life. I spoke with a professor of chemistry in his mid-sixties, for example, who referenced his background as "an old practical farmer out of Nebraska" as "where [he wants] to be engaged,"[9] indicating his desire to always have a connection to work with practical benefits. Socialization in a setting such as a farming community, which stresses practical benefits, may lead one to look past boundaries that existed between academic and commercially oriented work. Consider my conversation with a professor of bioengineering in his sixties regarding his turning point to commercialization:[10]

> My first time I filed a patent, a senior professor told me that that was not the done thing, that you should not file patents, that isn't what academics did. But I didn't grow up in an academic society. I grew up in a farming society, and I understood a couple things. I said, "If I have something that's of value, how is it going to get out there?" I understood the basic way that business works in America, and so I paid no attention to the academic stuff that was going around.

Indeed, he filed his first patent in 1976, when patenting was administratively difficult to accomplish and otherwise discouraged, even among engineers, as his account suggests. In this scientist's case, it appears that broader socialization in his life course outweighed the scientific norms of the era he discussed.

Notably, appealing to personal socioeconomic origins is found almost exclusively among commercialists who underwent professional socialization prior to Bayh-Dole. The explanation for this lies in the era in which they were trained, when commercialization was not just unconventional but shunned. These senior commercialists' appeals to having a long-standing disposition toward utility either vindicate their early departures from scientific convention or legitimate a "suppressed self," thereby permitting a deviant identity to "come out of the closet." By claiming a long-standing affinity to a practice now deemed legitimate by many in the community, these individuals are able to position themselves as moral entrepreneurs whose values are now celebrated. Their accounts also indicate the role sometimes played by nonscientific reference groups—that is, scientists are exposed to alternative motives for work through contexts of socialization other than academe.

Social Serendipity: "A Confluence of Events"

Science abounds with stories of serendipity in which accidents or chance coincidences are transformed into important discoveries. Commercialists among all the cohorts I interviewed regularly framed the circumstances leading to their initial commercial experiences as "a confluence of events" in which unexpected social interactions facilitated commercial turns. While they rhetorically emphasize the convergence of specific actors in space and time in a way that produced an opportunity structure for commercialization, other environmental properties clearly factored in to the onset of commercial activities, including the field and status of the scientist, university prestige, and the state of scientific knowledge.

I found this characterization of the commercial turning point more frequently among the scientists who received their PhD before 1980 than among the cohorts who trained after Bayh-Dole. In fact, several scientists within that senior group either formed their first company or applied for their first patent before the Bayh-Dole Act had been passed. Their status as "pioneers" of commercialization is partially a function of their expertise and partially the result of their location at premier universities, earlier choices of specific fields of study, and the advance of knowledge within those fields. The exemplar of this circumstance of commercialization is found in the 1970s, when the potential for applications and technologies resulting from developments in molecular biology were apparent to scientists in academe and industry. I spoke to a sixty-one-year-old professor of electrical engineering whose research has had medical applications.[11] When I asked him to describe his first patent, he said:

> Somebody knocked on the door and said "I understand you're an expert on this. Can you help me with that?" So we invented at the time a device for measuring [a physiological process] . . . The guy came in, and we sat down in here, came up with the idea of how we can do this, and we got a couple of patents on that. We started the company because he had this urge to commercialize it. Then I got involved with another company at the same time that funded us. So in a way, I got involved in two companies instantly with one guy walking in the door and solving a problem . . . He had gotten his PhD here maybe five years before, and he asked his advisor, "Who should I talk to about something like this?" and they told him go talk to this guy. I've had so many things like this happen over a lunch. I had a friend who was a neurologist. We had lunch one day with another guy who was an optics guy, and we filed four patents after that lunch, and that became a company. These things happen all the time.

In this account and the others it represents, the stage of scientific knowledge, the scientists' relevant expertise, their visibility within the scientific community, and the prestige of their institutions all precipitated the phone calls or events that triggered the initial commercial involvement. It is thus not exclusively a "lucky event" as some of these scientists suggest.

There is substantial overlap among the scientists who framed their involvement in this respect and those who stressed the practical aspects of their social origins. The scientists who were involved in commercialization prior to 1980 thus "ended up" as commercialists through a combination of external triggers and personal dispositions, yet they do not frame this transition in instrumental, active terms. Those who embraced commercialization more recently, however, frame their transition as circumstantial while also exhibiting intent. A professor of chemistry in his early fifties,[12] for example, told me about his commercial turning point as follows:

> So that was a confluence of two people walking down the hallway in opposite directions. It really was. I had been actively trying to transform my research program to one that was focusing more on renewable energy. My work has long been in the area of nanotechnology, but I didn't quite see what I'll call the "killer app" for what we were doing . . . And then suddenly one day, down the hallway comes this guy who is now my collaborator in this company, and he just stopped me and asked me if I knew anybody who could do electrochemical measurements. And I simply replied, "Well we're doing them all the time. Let's talk."

A Mertonian interpretation of commercialists' attributions of commercialization to "lucky circumstances" suggests the institutional norm of humility. This norm leads scientists to place limits on their accomplishments, in essence claiming only limited credit for their actions.[13] Particularly for those who were socialized to traditional means of achievement, citing a confluence of events allows them to claim only limited "blame" for an action that some view as morally questionable. It conveys that choosing to commercialize was a secondary consideration or unintended consequence, rather than a primary motivation or chief objective. But this interpretation, and also the scientists' accounts, obscures the operation of inequality in science. Sociologists Kjersten Bunker Whittington and Laurel Smith-Doerr have shown across different studies that female scientists engage in commercialization less than men but produce commercial work of similar or better impact.[14] Recent research has also shown that women were excluded in early interactions between academe and industry, which left female scientists with fewer opportunities and lower socialization to commercial science.[15] These studies tell us that although initial turning points may be experienced as a "lucky confluence of events," some scientists are privy to getting involved while others are not for reasons that may be related to inequality.

Whereas facilitators are preexisting or environmental conditions that ease scientists through the commercial turning point, attractors are a component of the evaluative framework within which scientists select between commercialist and traditionalist commitments. Here we turn to four attractors that lured scientists into the commercial role: material benefits, ethical boundary work, the potential for societal impact, and tangibility.

Material Benefits: "Money for the Lab"

Material benefits are an institutionalized component of a commercially oriented reward system. They are organizationally based rewards present at each of the universities that incentivize the commercialization of research with financial returns. Broadly, the material benefits derived from commercialization may be distributed to scientists, their universities (including the scientist's department, the technology transfer unit, and other administrative units), or organizations within the private sector. Among these, there are two forms of material benefit that directly impact scientists: personal remuneration and funds for research. Both signify an element of calculability because they are intertwined with quantification.

When I asked commercialists about the factors that influenced them to commercialize their work, they emphasized personal material benefits only rarely. Typically, when money was mentioned, its role was downplayed and the amount was dismissed as minimal. A professor of chemistry's discussion of licensing a patent to an energy company depicts how many commercialists discussed the influence of money:[16]

> My primary motivation was to get money out of the company to fund a research grant that I could get for students in the lab doing more chemistry . . . Just to be mercenary about it, we were able to use the fact that they were very interested in licensing this technology to leverage a research grant out of it simultaneously with the licensing contract. That's the only financial incentive for me to patent anything. It's not my major incentive . . . I put in a new bathroom in my house off of some of the licensing income, but it doesn't affect the way I operate. I'm not in this business to make money.

Commercialists tended to emphasize the financial benefits that return to their laboratories while downplaying personal remuneration. Academic science is often defined as under-rewarded and as fulfilling a social role in which money and prestige should be minimal or absent as a motive. Because university scientists have traditionally been characterized as disinterested experts who are committed to their work for its intrinsic value, a normative pressure limits the extent to which scientists would speak of the actual role played by material rewards. The commercialists I interviewed also frequently stressed the high probability of failure in commercial endeavors. This could mean that they viewed substantive financial gain as unlikely and thus material benefits were of limited influence, but this is improbable. It is not unusual for rarely attained rewards to motivate extreme commitment in science.[17] Thus, even if unlikely, the goal of "hitting the jackpot" financially could be operative in scientists' embrace of commercial activities even if it is not expressed.

Commercialists who entered science before the Bayh-Dole act generally dismissed the attraction of material benefits: they suggested that personal income and access to resources for research played little or no role in their decision to commercialize their work. This position was less prominent among commercialists who started their careers after Bayh-Dole. This subset of commercialists generally characterized money as a legitimate motivation for commercial involvement. The tone underlying these comments ranged from casual to emphatic. For example, a professor of biological engineering framed his embrace of commercial practices as following established precedent at his university,[18] an institution that had provided one of the earliest models of technology transfer:

> The patenting aspect was just something people do around here. We have a technology licensing office which returns 20 percent to the inventors, and with a young family that's attractive, and it also returns funds to the graduate school. So I just started patenting my ideas as they came along.

This engineer does not deny the influence of money, but he depicts patenting and profiting as conventional or "business as usual" without acknowledging that the magnitude of the royalties he earns is not easily dismissed. In the past year, his income from patenting, consulting, and corporate salary alone was approximately $85,000, more than half the average salary of a full professor at his university.

Other commercialists were more straightforward. Roger,[19] the biologist who leads the corporate-funded research institute depicted in the last chapter did not hesitate to mention money as a motive in forming his first company, which occurred largely in response to the development of gene-sequencing technologies.

This story, like a handful of other accounts of company formation, departs from the "typical" story of a company, which begins in a lab with a discovery that has apparent practical value and which follows a path to the marketplace. Instead, it emphasizes the desire to quantify scientific impact:

> The data started pouring out and then some friends and I simply started thinking, "Well, what can we do with that data stream?" We realized that most of it was gibberish, but we called around and asked whether anybody would give us money to set up a company to explore the utility of that data stream, and we were able to raise a large amount of money in a short period of time. That's how we got the first one started, and it's been quite successful. I made a lot of money, and all my colleagues and all the scientists we hired all made a lot of money.
> *What were you hoping to accomplish in starting this company?*
> I think we were interested in exploring whether we could do something useful, but we were certainly interested in whether we could make a lot of money.

When we examine the material benefits that arise from commercialization in a single year, the potential influence of money is not easily dismissed. On average, the commercialists I interviewed earned about $88,000 from patent royalties, consulting fees, and income from corporate positions combined. This was generally equal to half of their nine-month academic salaries, which range between $100,000 and $200,000. Over one-third of the commercialists I interviewed had commercial earnings that were equal to or more than their academic salaries.

In addition to these earnings, at the time of their interview, three-fourths of the commercialists had equity or an ownership interest in companies they had founded, licensed a patent to, or served as an officer or consultant. One-third of the commercialists had equity in two or more companies. Some of these scientists suggested that forming a company and selling it to a larger company is the best route to financial earnings. Indeed, a professor of biology I interviewed earned $40 million as a result of having 10 percent ownership of a company he cofounded and then sold for $400 million.[20]

While material benefits may indeed not be influential attractors in considering the pursuit of commercial endeavors, as many of these commercialists reported, still the commercial income they disclosed suggests that the magnitude of material benefits could be difficult to overlook. Moreover, recalling that, on average, over half of each commercialist's patents are the basis of at least one licensing agreement, it is difficult to conclude that such incentives are not learned and affirmed over time.[21] If material benefits are in fact influential attractors, it is likely that this view is disclosed only in circumstances of extreme trust. In short, social desirability, scientific norms, and the data on commercial returns suggest that the limited disclosure of financial motivations in these scientists' accounts does not fully represent the extent to which money encourages or sustains commercial involvement.

Ethical Boundary Work

Some commercialists in the study, independent of cohort, stressed the notion of ethical boundary work when explaining their initial entry into a commercial career path. By ethical boundary work, I am referring to the ways scientists negotiate circumstances in which credibility or intellectual territory is challenged in the course of scientific work.[22] Put simply, boundary work refers to how groups construct symbolic and social boundaries between categories, such as demarcating science from religion. Ethical boundary work entails identifying the line that separates ethical from unethical conduct, a border that is often ambiguous.[23] This mode of boundary work is a circumstantial attractor, in that the decision to commercialize one's work is part of an attempt to occupy a positive ethical space when confronting a complicated moral scenario. As we will see, the commercial trajectory attracts scientists because it offers a way of maintaining an ethical boundary that buffers threats to one's credibility.

Commercialists described two scenarios requiring ethical boundary work that prompted them to turn to commercialization. One scenario was the need to defend their claim to priority in discovery from other individuals or organizations who stood to benefit from their work. This mode of ethical boundary work typically occurred during consulting work that resulted in an unpremeditated commercializable technology, and commercialization was a way to protect oneself from misappropriated symbolic recognition or potential misuse of one's work. Consider, for example, the following account of a commercialist professor of chemistry in his early fifties who stressed a protective boundary when explaining his first patent:[24]

> I started working with a company that was trying to solve a problem, but in fact, they really only had about an 85 percent solution. The other 15 percent was what I knew better than they did. I worked to develop this widget. To protect myself, I decided, once I got this working, it was clearly superior to what they had, and I thought I should patent this. I just wanted to be sure that, you know, if they were relying on my knowledge that I was protecting myself. So it wasn't so much about the money. It was more about the principle.

Framed in such terms, boundary work functions as a symbolic counterpart to material benefit. Sustained commitment to traditionalism in this context would offer the disadvantageous circumstance of misallocated credit. Thus, by ensuring that credit—symbolic or material—is distributed fairly, protecting one's claim to priority attracts some to commercialization.

In the second type of scenario, scientists drew ethical boundaries in reaction to deceit or dishonesty. During the emergence of the biotechnology industry, for example, one commercialist professor of engineering made a discovery that had obvious medical applications.[25] When I asked him about his first patent, he told the following story, which occurred during the 1970s, when he was an assistant professor:

> There were some things that happened that led me to believe that the biotech area was going to be full of sharks and people who were willing to steal, including here on the faculty. A senior faculty member who was tangentially involved in [this field] came to me and specifically stated that he heard about [my discovery] and that he felt that it would be best if we sold this off as personal consulting to [a pharmaceutical company] and that he would set it up, because he was a buddy with a guy who headed [the pharmaceutical company's] vaccines. As soon as I heard that, I knew what that was. That was called theft, and so I just dropped anchor, that was it. As soon as a full professor at [this university] suggested stealing from it for a personal benefit, I knew my only ethical pathway was to get the patents free and clear, belonging to [the university] so that [it] could decide what to do with them.

In this example, ethical boundary work offers symbolic cleanliness. A paradox exists, however. On the one hand, entering a commercial career path could taint the reputation of a scientist, particularly in a cultural context in which commercial practices are viewed as morally questionable. Without disclosure, on the other hand, the scientist potentially fails to protect a discovery from the "dirtiness" of misappropriation, misuse, and professional misconduct. Formal commercialization of one's discovery therefore offers the legal and ethical safety of institutional protection, which may be viewed as advantageous relative to cultural sanctions or disapproval resulting from commercialization. The desire for protection thus nudges some scientists into commercial career paths by preserving symbolic rewards and cleanliness.

A third, less prominent scenario is deciding to patent one's research in order to protect it by securing it in the public realm. In this context, scientists elect to patent their research to provide free access to actors who would not profit from the use of discoveries while charging actors with a profit motive. Only one scientist I interviewed, a professor of biological engineering in his late sixties,[26] emphasized this form of ethical boundary work, noting that he has made materials available to other university researchers without seeking royalties while charging companies for access to the same materials. However, he did not discuss this form of protection as a motivation for his initial commercial turning point, nor did it emerge as a pattern among other commercialists.

Societal Impact

As we observed in chapter 2, commercialists view societal impact as both the most esteemed pursuit in academic science and a key basis for visibility, so it is not surprising that societal impact functioned as an attractor in their accounts of turning to commercialization. Societal impact as an attractor is tied to a craft ethic, or the impulse to manipulate and invent. As sociologist Richard Sennett argues in his work on craftsmanship, people invest in things they think they can change.[27] Societal impact in this context is a characteristic of the fruits of one's labor, separate from the desire to achieve visibility. Many of the commercialists who entered science prior to the Bayh-Dole Act set out to change the world through making contributions to knowledge, not through invention. They spoke of the technical and practical utility of their research as serendipitous societal impact. I spoke with a professor of biology in his mid-seventies who had patented almost fifty discoveries.[28] He described his first patent, a set of reagents, as "pretty fantastic" and said that his motivation was

> basically to make these reagents generally available. I mean, we only published, I think, a grand total of two papers and a review describing this stuff. As a matter of fact I [was] just at lunch with my co-inventor yesterday, and we [were discussing how] the amount of recognition we got for this was so close to zero.

Societal impact is framed here as a discovery unexpectedly uncovered in the course of research or else as an old solution to an emergent or new societal problem. This presentation of societal impact as a motivation mirrors explanations found in federal funding agendas, university mission statements, and the Bayh-Dole Act itself of why scientists should commercialize their work. It also approximates the notion of altruism, in that there is a sense of obligation to make inventions available to others to use in achieving their own objectives. That is, it suggests that invention is secondary to scientific discovery, not a motivation for it.

Tangibility

Scientists I interviewed who began their appointments in the 1980s or later sometimes referenced a variant notion of societal impact that reflects a different dimension of the craft ethic: the idea that invention is itself an impulse, not an unanticipated outcome of basic scientific research. Consider how a professor of materials science who completed his PhD in 1986 emphasized the material form of the first discovery he patented:

> The first project I worked on is one of the three most common [process] technologies used throughout the world, and these machines are sold all over the world, and I saw that happen and you know, I visit companies now . . . and they don't know that I had anything to do with it, and I love that . . . Here's these machines used all over the world, and that started on my white board in my conference room. I just love telling my kids that. I mean, they're proud of it. It's just cool because people found this useful . . . I've got over 200 publications, and it's very important for students to publish and things like that, but I don't need any more publications. Yes, it's great that people read about my ideas, but, you know, it's not enough to me.

What is evident in this scientist's account is the need to distinguish societal impact from scientific impact. The operative attractor at work here is tangibility of the invention. The attraction of commercialization for these scientists is the ability to perceive their work as materially existent, or capable of being realized beyond the status of merely a discovered truth. Rose, a chemical engineer in her early forties we have frequently heard from,[29] explicitly emphasized tangibility when discussing what led her to commercialize her work. For her, as for the scientist quoted just above, tangibility is what distinguishes commercial from traditional scientific rewards.

> I think every scientist that has an innovative idea wants to see that idea brought to a level of fruition that's satisfied. There is a different kind of excitement and a different kind of satisfaction that come from not only influencing your fields of study but also influencing what is happening in the everyday world. That's something that can only happen if there's commercialization.
> *You said there's a different kind of feeling. What's different about it?*
> I think it becomes more tangible. It becomes something that you can now explain to your grandmother or your kid. You see that do-hickey over there? That's because [of my work]. And I can actually say that.

Tangibility is what Sennett refers to as the dimension of presence in material consciousness, a scientist's quantifiable "maker's mark" that says, "I am here, in this invention."[30] Tangibility drives the intrinsic value of commercialization as perceived by commercialists who entered science after the Bayh-Dole Act. It explains why these cohorts of commercialists view commercialization as a legitimate end in itself, rather than an unanticipated consequence of their work. The rewards that tangibility offers, and thus its function as an attractor, include both the aesthetic value of materiality and the provision of a basis for distinguishing one's work from traditional scientific work. In contrast to scientific knowledge, which "goes into the ether," tangible inventions are concrete and quantifiable in market terms. Like commercialists' construction of visibility, tangibility evokes the element of calculability, a way to quantify societal impact that marks it as different from, and superior to, purely scientific impact.

The primary difference between societal impact as an attractor for senior cohorts of scientists and tangibility as an attractor for the more recent cohorts rests in whether or not these scientists anticipate the materialization of a product from research. Compare, for example, the above expressions of the satisfaction derived from the tangibility of invention with a senior biologist's description of his reaction to seeing his first invention materialize:[31]

> What I never expected was within twelve months to come into a major national hospital and walk into their clinical lab and see it running. I couldn't believe it, frankly. It worried me [because] they are relying on this 100 percent, and now it translates into lives. It's no longer academic.

Whereas more recent cohorts of commercialists envision an invention and are motivated by the materialization of their work beyond academe, this kind of tangibility was unanticipated by commercialists whose training preceded the acceleration of commercialism in higher education.

The analysis of attractors and facilitators of commercialization among the commercialists points to divergent patterns between the scientists trained before and trained after 1980. The general tendency of commercialists trained before the acceleration of commercialization is to present their initial entry into commercial career trajectories as indirect or not their central motivation. They frame their early experiences as resulting from the serendipity of research, the circumstances of their early lives, or social serendipity. They rarely present their transition to commercialism as sought out or as motivated by the institutionalized financial goals that result from legislation. These patterns overlap somewhat with commercialists trained during the 1980s and 1990s, although the later cohorts show a more pronounced identification with commercialization as an intrinsically valuable activity. Facilitators push scientists toward and into the commercial career trajectory, while material benefits, ethical boundary work, societal impact, and tangibility attract scientists to commercialization because of the perceived advantages they offer.

Why Do Traditionalists Remain Traditionalists?

If commercialization has radically accelerated in higher education, and if commercial culture is pervasive at elite research universities like those of the scientists I interviewed, why are traditionalists not persuaded to embrace a commercial turn? The answer is not for want of interaction with industry. Some of the traditionalists I interviewed have been courted by industry. Some even flirted with the idea and took meetings with companies, but ultimately remained steadfast along the traditionalist path. The findings I present in this section indicate that explanations related to differential opportunity structures fail to explain the lack of commercialization among the traditionalists, who are exposed to the same incentives and opportunities as commercialists but are untempted or "repelled" by commercialist pursuits. The discussion is organized around a consideration of five social mechanisms that constrain identification with the commercial role: professional socialization, opportunity structure, disinterestedness, career burden, and goal incompatibility.

Professional Socialization

Traditionalists who trained before 1980, like their commercialist peers from the same cohorts, were well insulated from early commercial developments in higher education during their doctoral and postdoctoral training. They also used equally colorful language as their commercialist peers to characterize the view of commercialization that they were exposed to during their professional socialization. Even scientists who trained at places like Stanford or Caltech, which have long-standing traditions of industrial relations, noted that they were exposed to a negative view of commercial practices during this era.[32] Ruth, a traditionalist we have frequently heard from and who completed her doctorate well before the Bayh-Dole Act, told me that there was at Caltech then very much "a spirit of science is a pure enterprise that you do because it's there, and you don't do it to get rich."

Whereas commercialists whose training coincided with or followed Bayh-Dole viewed commercialization as elemental to the scientific role, traditionalists from these cohorts generally developed an unfavorable view of commercialization during doctoral training, as their exposure to commercialization entailed cautionary tales and examples of career outcomes to avoid. For some, close observation of the experiences of commercialist advisors exposed them to negative aspects of the commercial role, leading them to reject commercial pursuits and affirm traditional science. Five of the traditionalists trained under a commercialist during graduate or postdoctoral training. An associate professor of chemistry said about his first exposure to commercialism:[33]

> My advisor at Berkeley has a company. I was able to observe things that he chose to do that he lamented, would complain about, or did not seem to actually enjoy. He and I were similar, we got along quite well, and so I was able to gauge that if he didn't like something, I was probably not going to like that same thing, and he never really seemed to like being in charge of [his company].

For scientists who chose the traditionalist path after exposure to downsides of commercialization, commercialization is viewed as a consequential distraction from traditional means of achievement. Dissatisfaction, opportunity costs, and negligence in one's role become associated with commercial behavior, outweighing whatever value is assigned to commercial incentives. These concerns do not universally produce a view of commercialization as morally questionable, but they do affirm traditionalists' identification with and commitment to the conventional means of achievement and reward by minimizing the potential for positive identification with the commercial role.

Some traditionalists I interviewed did develop negative conceptions of commercialization during their training through adopting their advisor's view, observing commercialization as a constraint, and witnessing negative developments such as litigation between universities and commercial firms. An assistant professor of biology,[34] for example, described how a nine-year patent litigation dispute between the University of California San Francisco and Genentech was "certainly noticeable" during her doctoral training at Berkeley. She characterized her advisor as "annoyed by seminars partnered with Bay Area venture capitalists" and said that her advisor "hated what she called all the 'whispering in the hallways.'" I asked her if she had ever considered a commercial career path in academe, and she replied:

> No, no. And quite notably not, because my very close friend and bay-mate, the guy that I worked with my back two feet from for all of graduate school, was very interested in industry and the private sector and he participated in a couple of

> "idea to IPO" type seminar class–based competitions, and it seemed like a distraction. It worked out better for me to focus on my PhD research.

Traditionalists who are trained by scientists with negative views of commercialization do not develop expectations of commercial behavior, largely because they emulate the scientific role as it is enacted by their advisors. Socialization in such a context, coupled with exposure to pitfalls, distractions, and paths that failed to pay off in economic or scientific currencies for commercialist scientists in the same department, leads to conceptions of commercialization as a career threat—despite commercialism's presence in the form of commercial peers, federal agendas for technology transfer, and institutional encouragement of innovation.

Opportunity Structure

One possible reason that traditionalists do not attempt to commercialize their work is that their work contexts are not characterized by opportunity structures conducive to commercialization.[35] Given the theoretic framework of the study's sample—in particular, the identification of commercially prolific institutions, departments, and scientists and the selection of traditionalists who work in the same environments—it is difficult to conclude that these organizational characteristics explain why the traditionalists in this study do not commercialize their work. Indeed, given their regularity of interaction with commercialists, access to the same technology licensing resources, and to some extent coauthorship with commercialists, one could argue that the traditionalists I interviewed have a greater opportunity to "learn" commercialization than most other scientists in the United States.

Furthermore, the research foci of the majority of the traditionalists I spoke with were not irrelevant to technological or commercial concerns. Only two of the scientists I interviewed explicitly claimed that no ties existed between their research and the potential for technology transfer. The career of one of these scientists suggests that in some cases that perception may be selective. For example, earlier in the interview that scientist noted that industrial scientists have cited his work. Additionally, he noted that he has consulted for four companies over the course of his career. The purpose of this illustration is not to suggest that most scientists could commercialize their work but for lack of desire. Rather, the point is that among some scientists, commitment to the study of fundamental problems renders any consideration of practical applications cognitively unrealistic.

Disinterestedness

Most of the traditionalists I interviewed told me that they are unpersuaded by commercial opportunities because they are interested only in the discovery of new knowledge, not in its application. As a result, many traditionalists forego the opportunity to pursue potential applications of their research. Consider, for instance, a traditionalist professor of chemical engineering in his early sixties who,[36] when asked about the extent to which he had considered or sought to apply his work commercially, stated:

> In some areas where I've worked there is definite industrial interest. Many companies make membranes for things like desalination or purifying protein solutions or other kinds of applications, and so various times I've been a consultant with those companies helping understand what their device was doing and sometimes how to improve it, but that's as close as I've gotten.
> *Given these connections, does commercialization not attract you?*
> It just doesn't interest me that much. The commercial side of things doesn't interest me that much.

I pressed this engineer to imagine circumstances under which he would commercialize his work, but he could not, commenting that such a scenario would be "extremely hypothetical."

A department of engineering at a commercially intensive university is the type of environment in which high levels of adherence to commercialism can be expected, yet even among researchers in such a department who are engaged in work with "definite industrial interest," we can still find adherence to the norm of disinterestedness. Traditionalists' accounts implicitly reflect a commitment to selecting research problems motivated by scientific rather than nonscientific concerns. More generally, this reveals their commitment to the institutional goals of science, since their identification with the goal of advancing knowledge is unperturbed by potential career paths that depart from that objective.

Career Burden

Some of the traditionalists I interviewed suggested that they were not interested in commercialization because it would interfere with their career, in particular their time and reputation. In chapter 2, we see that one reason traditionalists are critical of commercialists is that they see commercial activities as detracting from commitments to graduate students and departmental responsibilities. A similar narrative emerged when traditionalists were asked if they could envision a commercial career path in connection with their own research. I spoke with a professor of biology in his early sixties,[37] known for important discoveries in genetics, who, when asked whether he had considered commercial activities related to his work, replied:

> I've never had the time to even think about it. I mean, I've always been so struggling to keep up . . . I got so far behind on publishing work that we've done that I think that was also one reason why . . . I found it so hard just to keep up with my own area and my own line and my students and postdocs, that I've really never had a chance to think about that.

This mode of aversion to commercial behavior lacks negative attribution. It simply points out that in university environments where scientists must juggle competing demands, any possible commitment not highly valued—in this case, commercialization—is likely to be avoided. In this respect, commercialization is constructed as an undue burden, the weight of which overshadows any potential benefits one might derive from it.

Other traditionalists who suggested they could never imagine a commercial career path told me that they saw it as too risky to their careers. Another scientist,[38] who has closely observed the commercial career paths of his colleagues since joining a commercially intensive biochemistry department in the early 1970s, explained:

> It's never occurred to me to do that, ever. You have to convince people that what you're doing is really very promising. Most of us are trained, as scientists, to be very skeptical about things, you know. You look at your own results very critically, and you interpret them as conservatively as you can so you don't get caught out on a limb and have to be embarrassed by taking it back, and so on. When you go out to get a millionaire and go back to your small startup company, you can't say, well, you know, it may not amount to anything . . . The last thing you want to have happen is you getting caught having made a claim that's not justified. It can be the end of your career. You can keep doing science, but people take you less seriously if you've ever done that.

I learned later in another interview that a commercialist in this biologist's department had been accused in a patent lawsuit of fraudulent claims to a discovery underlying a patented invention.[39] Although the case was settled out of court, and three independent investigations cleared the commercialist of any wrongdoing, such events call scientists' reputation and credibility into question. Scientists may be exposed to both incentives and sanctions for commercial behavior. Given the years invested in establishing one's visibility within a field, the possibility that commercialization could damage one's reputation may persuade scientists who have observed such outcomes to avoid the commercial career path.

Goal Incompatibility

Some of the traditionalists I interviewed had either been approached by industry or had approached the technology transfer offices at their universities at some point in their career. But in the end, for these scientists, adherence to basic research prevailed—often due to perceived inconsistencies in the goals or interests of the industrial scientists and representatives of industry with whom they met. Most of these traditionalists characterized their interactions with industry as brief and uninteresting. For example, I spoke with a professor of biology in her sixties who,[40] because of her leadership in a particular field, had served on the boards of two drug companies. She ultimately stopped, she said, because she "thought they were more interested in what they could patent then what they could find out." Most traditionalists in the study developed similarly unfavorable definitions of commercialization as a result of this type of commercial encounter, characterizing the experience as "just a waste of time for me,"[41] "a complete disaster from the beginning,"[42] or "just boring."[43] Comments by other traditionalist interviewees suggested that traditionalists view the commercial problems as unchallenging, the problems as short term, and the goals as unrelated to knowledge.

Even when there are relatively direct connections between their research focus and a company's interests, traditionalists appear unmotivated to pursue commercialization, although they may be willing to share tacit knowledge related to a problem. I interviewed a professor of chemistry who discussed research he had published in *Nature* that attracted interest from industry.[44] Compared with most traditionalists in the sample, he had a generally neutral or open view of commercialization, but he nevertheless avoided the opportunities that became available because of his expertise. He said regarding the one time he gave commercialization "any serious thought at all":

> We had done some experiments that got a bit of press, and somebody from a company approached me. I and a postdoc went over to this company and visited with them. It was clear the piece of our work they had read about wouldn't work [for their purposes]. We saw that from the get-go. There was another thing we had done that might work, and so we showed them that approach. They got

really excited about this and they were saying . . . if you could [do this] the world would beat a path to your door. This company got really excited. I could have tried to build a relationship with them and put a lot of my effort into tuning up this process, but I just didn't want to do that.

In some respects, the traditionalists' commercial encounters reflect the orientation to commercialization they feel their commercialist peers should practice: commercial trajectories should be episodic events rather than definitive attributes of an academic career. The interviews suggest that traditionalists are open to or support the dissemination of practical results—otherwise they would not entertain interaction with industry—but they are unwilling to commit themselves to this end if doing so interferes with the commitment to fundamental research.

Only one traditionalist I interviewed, a professor of biology in his early sixties,[45] has a sustained consulting relationship with a company, for which he received between $80,000 and $90,000 a year in fees. Despite the advisory role he played and the obvious connection between his research and technological or commercial opportunities, he remained uninterested in conducting any form of commercially oriented research. I asked if he had considered commercializing his work at any point in his career. He answered no and continued:

I was a consultant for a company called ["LifeSci Inc."] when it first started, and I actually made a lot of money on that. They gave me a stock option, I did well by it, and I advised them on their practical goals. But I didn't feel that I wanted to do that in my own lab . . . The entire world's supply of [a virus] vaccine is made [following a process I developed] and I wouldn't have tried that in my own lab, but to see it happen with my advice was exhilarating.

Traditionalists view sustained commercial commitments as not worthwhile investments of time and as neither interesting nor attractive. Although they work in areas of science that have obvious connections to commercial technologies and have similar opportunity structures as their commercialist peers, for them, adherence to the goals of science precludes identification with the commercial role. In general, traditionalists appear to be repelled by, rather than attracted to, commercial involvement. Any flirtation with the notion of commercial involvement is neutralized with the perception of contradictory interests between university and industrial science. In short, for traditionalists, opportunities to commercialize research lead to an affirmation of the goal of advancing knowledge and the traditional means of achievement.

Conclusion

Consideration of the processes by which scientists are socialized to commercialization and the factors that lead some scientists to embrace commercialism and others to maintain career paths oriented around traditionalism reveals that professional socialization during doctoral and postdoctoral training has a huge influence. Among commercialists whose training preceded the Bayh-Dole Act, the embrace of commercialism often occurred in late career stages, in part because of the intensive socialization to traditionalism they had received. Yet for more recent cohorts of commercialists, commercialism was an elemental component of their training. It is also clear that embeddedness in highly commercial contexts is a critical factor, even if commercialists attribute their commercial trajectories to social serendipity. Money, ethics, the desire for impact, and an occupational aesthetic of tangibility also nudge scientists toward commercialism. Traditionalists espouse commitment to the conventional goal of advancing knowledge to the extent that opportunities to engage in behaviors that depart from this goal either affirm their commitment to science or are dismissed outright. We saw that many traditionalists have experienced one or more encounters with industry. Perhaps the rhetoric of commitment to traditionalism may be less important than how such encounters unfold.

To complete our understanding of why academic scientists embrace commercialism and traditionalism, we must not limit ourselves to the mechanisms that led faculty to one path or the other. We must also consider how they maintain their professional identities as commercialists and traditionalists in intensive contexts of academic capitalism, where competing ideologies of work are operative. Let us now examine patterns of identity work in which the scientists engage to construct and affirm a desired professional self.

CHAPTER FOUR

Identity Work in the Commercialized Academy

Roger, a professor of biology,[1] has started companies and now runs a research institute funded primarily with industrial money. In light of his involvement in commercialism, a comment he made during his interview struck me as remarkable. He said commercialization

> was unthinkable to me when I was young. I went through a very long period of time where even though I was offered money to work with companies, I mean research money in my lab as well as consulting money, I didn't take any money into my lab because I didn't want it to be *contaminated* by any kind of industrial money.

How do commercialists such as Roger, who were originally socialized to view commercialization as a dirty, contaminating thing, ultimately see themselves in a positive light? How do they attempt to secure the affirmations of others who might question the virtue of commercial activities? And how do traditionalists who work in commercially intensive environments cope with the identity threat commercial culture poses to traditionalism?

The answer to all these questions is identity work. Identity work refers to individuals' attempts to maintain a coherent self-identity in the context of shifting experiences.[2] Maintaining a coherent self-identity is particularly challenging as a result of the decreasing influence of tradition in society, because changes and disruptions in patterns of events destabilize basic ideas about the self and the relationship between the individual and society.[3] Change-oriented situations in science arguably lead to increased identity work for both commercialists and traditionalists, who develop interpretive schemes and concomitant self-identities to make sense of their place in a changing academic profession and thereby achieve their goals. Sociologist Paul DiMaggio notes that "collective identities are

TABLE 8
Identity Work among Commercialists and Traditionalists

	Commercialists	Traditionalists
Modal process	Legitimation	Disidentification
Techniques	Reframing	Disavowal
	Social comparison	Social comparison
	Role distancing	Ritual identification
	Professionalism	Retreatism

chronically contested" because groups within a community vie to produce interpretive schemes that are favorable to their own ideal or material interests.[4]

Through the interviews I conducted, I uncovered competing forms of identity work, provoked by commercialization, among traditionalists and commercialists. The objective of the identity work is predicated on the reality that both traditionalists and commercialists are aware of a moral taint surrounding the commercial role. The emergence of the commercial reward system and its perceived "dirtiness" has had a destabilizing effect on notions of identity in science. Through my interviews I identified two broad processes—legitimation among commercialists and disidentification among traditionalists—and two sets of four respective techniques by which those processes occur. The objective of these processes for the commercialists is negation of moral taint, and for traditionalists, the affirmation of purity (table 8).

Engaging in techniques of legitimation is particularly necessary for commercialists who were professionally socialized in contexts of ideological purity, in which they developed a view of the commercial role as dirty—that is, they earned their PhDs before the acceleration of commercialization in academe after 1980—but this is equally true for commercialists who entered science more recently. Even though some groups (such as the Association of University Technology Managers or university technology transfer offices) actively promote commercialization, there is nevertheless a "moral taint," or dubious virtue, associated with the commercial role, a view that is embraced by critics in and beyond the academic community who believe there are morally questionable aspects of commercialization. As we will see, commercialists engage in four techniques of legitimation to maintain a positive self-identity and to attempt to present a legitimate professional self to the scientific community. The result is that commercialists are able to engage in a confrontational, or "dirty," task while presenting a professional identity that is "clean."

Identity work among traditionalists is motivated by the destabilization of the traditional academic role and the aura of taint associated with commercialization. Perhaps because of the pervasiveness of commercialization in the departments in which the traditionalists are situated, these scientists seek to affirm their own purity, even though they do not actually participate in commercial culture. Commercialization in this respect constitutes the basis of a "never identity" for traditionalists, as the referents of commercial culture are drastically inconsistent with what traditionalists prize. Traditionalists therefore engage in four techniques of disidentification, the goals of which are to affirm purity by repudiating commercial referents.

Commercialist Techniques of Legitimation: Negation of Moral Taint

Previous research in the social psychology of work suggests that the reasons and meanings of commercialization must be analyzed beyond their function as motivating factors. By identifying neutralization techniques that rationalize and legitimate inconsistent behavior, moral justifications for questionable conduct, or socially approved vocabularies that serve as explanatory mechanisms for deviance, social psychological studies frequently demonstrate how workers secure a positive identity in the context of morally questionable work.[5] I draw upon this approach here to demonstrate how scientists who patent and form companies legitimate their involvement in commercialization using four modes of identity work: reframing, role distancing, social comparison, and professionalism. Each practice, through varying means, seeks to negate moral taint or neutralize imputations that commercialization is harmful, thereby permitting a positive and morally appropriate presentation of self.

Reframing

Occupational ideologies are action-oriented beliefs held by members of an occupation to make sense of their work. As coherent belief systems, occupational ideologies specify role expectations and the nature of the relationship between members of a profession and society. Workers will engage in self-serving ideological techniques when work is characterized by conflicting interpretations, as in this study, in which some scientists view commercialization as contaminating while others view it as esteemed. Ideological techniques have their greatest impact on securing internal legitimacy because, by sharing beliefs within an occupation, members persuade other members to identify with the role in question.[6] Learning to legitimate commercialization thus plays a critical role in a scientist's socialization to, and persistence in, commercial involvement.

New commercialists learn from others how to legitimate their role through the process of reframing, whereby morally questionable practices are imbued with positive meaning. Reframing is necessary simply because the meaning of a role always differs depending on whether one is in the role or is on the outside looking in; hence the necessity of normalizing a work setting when one enters a new career path. Even a scientist socialized to a favorable view of commercialization will ultimately reframe its meaning because, prior to transitioning to the role, occupational myths belie what one learns through experience.

Reframing is also necessary when a stigma must be overcome in order to embrace a role. I encountered this mode of identity management when I interviewed scientists whose first commercial act involved company formation. I spoke with a gregarious forty-year-old associate professor of chemistry who was in the process of forming his first company when we met.[7] He described himself as "first and foremost a basic scientist interested in fundamental questions." This interview was memorable because he was apparently quite torn about his new commercial trajectory—a position he referred to as "devil and angel," reflecting his excitement and reservation. During the year preceding the interview, in collaboration with a colleague and a doctoral student, he had filed for and received his first patent and formed his first company. He mentioned this even before I asked him about commercialization as we discussed the role of the academic scientist, and he began to explain his feelings:

> I never had any intention of patenting anything. What triggered us down this pathway is that a company got in touch with us after reading one of our publications and was very much interested in the work. One thing led to another, and it made sense to patent the work. At that point I started to have to worry about, at what point can we start to share this information in the public domain? I will be honest, it bothers me . . . I think the way it was put to me by my colleague was very influential. He said, "From the standpoint of you as an advisor, you will learn things in starting a company that make you a better advisor." That was the selling point for me. I really struggled with this. Is this something I really want to do? And when it was put to me in those terms, that I think I'll come out of this whole process and be a better advisor for it, I was convinced it was the right move for me . . . I wrestle with this a lot. I still do even though we've obviously now taken the plunge, so to speak.

The ambivalence and apprehension that color this account are derived from the dissonance between a previously envisioned and a present self. As I found in other commercialists' descriptions of commercial turning points, these concerns are quelled by infusing aspects of commercialization considered to be problematic with positive meaning. In this case, this associate professor learned to reframe commercialization as a practice that will enhance his ability to train and guide graduate students. Legitimating the new role is particularly essential given the extent to which a new self-identity departs from what was forged during professional socialization. We can see this in the contrast between how he described his view of commercialization during graduate school and his present ambitions for a commercial future:

> [My doctoral advisor] and I would talk about these things, and we'd kind of hold our nose up and say . . . "That's not our role in this job" and "This is all about mountain-climbing and ego-boosting" . . . My sense coming out of grad school was, This is not what academics is all about. We shouldn't be doing this kind of thing. Our primary mission is to be teaching and worried more about that and less about financial reward, the private sector, and the application side. I see the commercialization much more as technology and much less about hypothesis-driven science. The academic in me has always been more driven by the questions of basic research . . . I could be accused of being hypocritical because I was pretty vocal about this ten years ago, just like my advisor was. "We're here to teach and do academic research, and our job is not to start companies."

His interpretation of commercialization thus involves a derogatory view in the past and a "devil and angel" or hypocritical view in the present. But consider how he reframes and constructs new expectations of a commercial future:

> The idea that I would understand anything about human physiology was so foreign to me a few years ago, and now I see myself more interested in it, and I think that there are some really exciting opportunities, and as I learn more about it, I discover that there are even more problems for me as a basic scientist—a basic scientist who can design and make molecules.
>
> *Now that this is starting to materialize, what have your thoughts been about making money?*
>
> When I daydream about this, I relish in the idea that I might have enough of a revenue stream coming in to support my research that I don't have to worry about the vagaries of the government funding system. Although I do think writ-

> ing NIH grants is good for the soul, I'd rather not be beholden to it. I'd rather be able to write my own checks for my research. That would be terrific. My wife and I do all right. We live in a very expensive part of the world, but we both work and we have a nice life. We don't live in a rich house, and we certainly don't have a lifestyle that requires a lot of money. Look at my office and how spartan it is. This is how my life is. This is the person I am. I don't think having 30 million in my bank account would probably change me in any way. A few more nice dinners out, but I don't think I'd drive a Tesla or anything of the sort. People laugh when I say to them that really my goal would be to make enough money that I wouldn't have to write another research grant. There is some truth to that. I think the other thing I would love to be able to do is to give money philanthropically. To have that kind of money to be able to help out a university like [this one], to give money to my alma maters, which I love dearly . . . I would love to be able to give lots of money to those departments to fund graduate fellowships or undergraduate fellowships. That would be extremely rewarding if I someday had that kind of money.

We can see how, across these excerpts from his interview, the professor shifts from being "first and foremost a basic scientist" to being one who can "design and make molecules." The addition of invention to the role set of the scientist has catalyzed new dreams for a career future that depart from the scientific goal of advancing knowledge. Future goals entail solving societal problems, but also lifestyle construction. Financial incentives evolve into fantasies of the future and lead in many directions, including professional autonomy, philanthropy, leisure, and material possessions.

Patterns of ambivalence and learning to legitimate are also found in the accounts of commercialists whose transition to commercially oriented careers had already occurred. Charles, a professor of chemistry we have frequently heard from,[8] was discussing his commercial goals when he told me a story about his first company. Notice how he has an initial aversion to embracing financial incentives that he then reframes:

> I did learn something from that first little company because I met with [another scientist] and he sat down with me. "So why do you want to start a company?" he asks. And I basically said, "Well, we're really interested in having another vehicle for doing research." The guy looks at me and says, "If nobody in this company is doing it to make money, it's probably not going to be successful." At that time I was still struggling with the moral issue of admitting to myself that a motive for making a company was to make money. I wasn't there

at that time. I really hadn't thought it through thoroughly, because if I would have, I would have thought that was dumb. By the time I started [my second company] that thought had matured in me, and so I now was interested in making money. It actually was a straightforward calculus. This invention was a big hit, like a home run slammed way out of the park, and made a ton of money for [the company], like billions.

What was the logic behind having a company to do research?

There was no logic behind that really. It was just because I wasn't willing to admit to myself that money is not what I . . . you should start a company for.

So it was a rationalization?

I think so.

Like the other chemists, Charles's account of committing to start his first company reflects a process in which an existing view of commercialization as morally tainted is transformed. Constrained by his adherence to traditional scientific norms, he had unsuccessfully attempted to enact an ideology of commercialization that was consonant with traditional science. The outcome of reframing is evident in the meaning he now assigns to commercialization. For example, consider a comment Charles made later in the interview regarding technology licensing offices:

> [They are] doing their job of going around and seeing what uncollected diamonds are lying on the floor. If there's a bunch of uncollected diamonds lying on the floor that someone is going to vacuum up and they'll be in the landfill, why not go pick them up? Commercialization is just a resource allocation issue by the university and the tech transfer office.

In this comment, Charles assigns a very different meaning to money than in his earlier comment. It has shifted from a questionable, illegitimate motive for action to a simple issue of resource allocation, without which opportunity costs would manifest.

These examples demonstrate how learning to legitimate the commercial role through reframing resolves cognitive dissonance.[9] Regardless of the context of their socialization, the scientists initially exhibited ambivalence, apprehension, or incomplete acceptance of the commercial role as a consequence of the conflict between their new activities and how they wanted to see themselves. Because the practices and norms that correspond to a commercialist self-concept depart from existing cultural expectations in academic science, new commercialists perceive their departure from traditional scientific convention as morally questionable.

The discrepancy in self-concept motivates behavioral and attitudinal change, the result of which is the internalization of meanings of commercialization that substantially depart from the views to which the scientists were socialized. Consequently, dominant mechanisms of social control are rendered inoperative. Reframing is thus a technique of identity work critical to transforming traditionalists into commercialists.[10]

To sustain this self-identity and persist in the commercial role, however, commercialists must learn other techniques to legitimate externally as they become known for, or associated with, commercialization. With the capacity to neutralize the negative associations of commercialization with royalties, companies, industrial affiliation, and commercial norms, a commercialist identity begins a course toward institutionalization in academic science.

Role Distancing

Rarely do expert workers or craftsmen seek to disassociate themselves from the fruits of their labor. Rather, classic depictions of expert labor emphasize both the extent to which workers see themselves in their products and the extent to which the inability to do so results in alienation.[11] Indeed, as I show in chapter 2, commercialists frame their visibility in terms of their ability to cite broad societal impact through their technological inventions. In circumstances in which the commercial role is called into question, by contrast, creating symbolic distance between one's sense of self and one's creations provides a means of minimizing one's investment in that role. The intention, of course, is to minimize one's association with morally tainted aspects of commercial involvement, such as the notion that commercialization threatens one's commitment to university responsibilities or the ability to buffer one's academic work from industrial influences.

This is a difficult feat, particularly because of the permanent association derived from property rights and the paucity of free time in a scientific career. Consider, for instance, my interview with a thirty-eight-year-old associate professor of chemistry at a public university who informed me at the time of the interview that "literally, we are forming [the company] as we speak."[12] As we discussed his motivations for starting the company, this chemist expressed reservation about time commitment and the management of conflicts of interest. When I asked him how he reconciles the tensions he described, he responded:

> I guess I've rationalized it as: It's not my company, it's my co-worker's company. I'm just kind of a glorified consultant. Okay, cofounder is next to my name.

> *Don't you have an equity interest?*
> Absolutely. There's no question that I have a real financial stake in this, and I put my own money into it. But I guess I just don't . . . in my mind, I just don't see it as my company. Truth is, the ideas for this company are largely out of my own brain. My collaborators across the street, I love them dearly, but I think they would be the first to tell you this was my electro-property and me coming to them saying . . . "Guys, I have something here, are you interested?"

The intellectual property underlying the endeavor, he acknowledges, is recognizably his, yet he symbolically resigns himself to the status of "glorified consultant." As a technique of neutralization, here role distancing may be serving both personal and public objectives. For the commercialist himself, rationalization may function as a means for sustaining a positive identity. With respect to others, it may form the basis of attempts to persuade the scientific community of his "cleanliness" by placing the "dirty work" at a distance.

Other scientists who I interviewed engaged in role distancing by suggesting that commercialization is a peripheral rather than a central component of their work. It was not uncommon, for example, for scientists to say that their involvement in commercial activities entailed a small portion of their time. They framed their participation as a "bonus," an "adjunct to basic academic life," "an interesting dimension to add to the scientific career," and "something done off salary," and thus not a threat to the performance of their university obligations. Other acts of role distancing were externally directed, targeting colleagues. Consider my interview with a Nobel laureate whose commercial activities resulted in past year revenues of $1 million for his university and more than $270,000 in personal commercial income.[13] However, he downplays or distances himself from these markers of commercial success in the way he presents himself to his colleagues:

> Clearly having more money in some ways has made my life easier to manage. I hire people to fix the house, I hire people to do things that I otherwise would have to spend my time doing . . . [But] I don't want to live in a way that shows money, because it would separate me from the colleagues that I value working with.

The visibility brought by an award such as the Nobel Prize stands in sharp contrast to the obscurity he attempts to cast over his commercialist attainments. That he engages in role distancing to avoid alienating his colleagues underscores the identity challenge and conflict provoked by having commercial status.

The most subtle use of role distancing is found in the omission of commercial activities from a curriculum vitae and professional webpage. More than one-third of the commercialists who founded companies at least one year before interviews occurred did not disclose this information on a CV.[14] This is striking because scientists tend to list their accomplishments extensively in these documents. As a historical map of one's achievements, experiences, and affiliations, a scientist's curriculum vitae provides the most detailed public account of the scientific career, from the momentous to the minute. To not disclose a product of one's work, particularly one that entails commitment of greater duration than, for example, a guest lecture or participation on a committee selecting a journal editor (types of activities I frequently found on scientists' vitae), is therefore suggestive of purposeful role distancing.

Social Comparison

Another way to neutralize the negative aspects of a social object is to selectively compare it to something similar that is deemed inferior—a downward social comparison. Doing so enhances the object in question by virtue of its position relative to a more disreputable, discredited, or morally tainted referent. This is a subjective portrayal, for it involves disparagement and attributions of inferiority in order to portray the object at hand in a more positive light.

Whereas participants in some dirty work occupations engage in selective social comparisons in reference to other occupations, commercialists' use of this technique involves comparing commercial activities in reference to other practices or regularities of academic life—an intra- rather than interoccupational comparison.[15] A common example, confined to commercialists at public universities, involves comparing commercialization of research with the publication of books and textbooks. This practice emerged most frequently when I asked the scientists about the impact of commercialization on collegiality. I interviewed a fifty-seven-year-old professor of biology at a university where there had been past episodes of tensions related to close industrial ties.[16] When I asked him whether he had colleagues opposed to commercialization, he responded:

> Probably in the humanities I would say . . . But some of these humanities professors don't hesitate to write a book. Commercialize a book. So what's the difference between that and a company, right? Same basic concept. The time spent to write a book versus starting a company has got to have some analogies there. That's more what I'm referring to. Not how much money . . . Whether it

> fails or is a success is one issue. But getting to that point where it's out there is another issue.

At minimum, selective comparison invokes the social norm of reciprocity, in that the legitimacy of commercialization is framed as "no worse" than publishing a book or textbook.[17] By engaging in this type of selective comparison, commercialists accord themselves the autonomy indicative of academic work. The moral taint associated with commercialism is negated by equating it to a long-standing academic tradition.

More frequently, however, commercialists disparaged book publishing as having little or no value. In an interview with a professor of biochemistry in her forties at a public university characterized by a long history of commercialization,[18] I asked if commercialization impacts collegiality, and she responded:

> I don't think so. One thing that you're not covering that is actually similar and no one questions is if I write an undergraduate textbook in organic chemistry, I can make a ton of money, and I can spend a lot of time doing that, and people view that as perfectly reasonable, scholarly work, but it's not forwarding the field. So if you compare that to entrepreneurship, I have to rate entrepreneurship ahead of writing a textbook. People update those all the time, and what it does is it just makes a lot of money for them and the publisher, and that's it.

Note that in both this and the preceding interview extract, the scientists avoided directly addressing the question posed to them. In contrast to reframing and role distancing, which constitute sustained modes of legitimation that begin once one enters the commercial role, social comparison represents a reactive technique. Theorists of social comparison argue that this technique is prevalent when focal actors experience threat.[19] Given that the question asks participants to reflect on potentially negative outcomes related to their commercial involvement, it "calls into question" the legitimacy of commercialization.[20] At public universities, where scientists who write textbooks may more likely be found than in private institutions, commercialists who find themselves "put on trial" have a foil at their disposal to present commercialization in a more positive light.

Commercialists utilized selective comparisons in other instances as well. Again in response to questions interpreted as threatening, scientists responded by comparing allegedly negative aspects of commercialization to traditional aspects of academic work. Discussing the patent review process, I asked one professor of biochemistry,[21] "Do you see it as problematic that patent reviewers are not experts in the areas you do research in?" He responded with a selective comparison:

> Well, there is a difficulty in getting patents through the [United States Patent and Trademark Office] because the people that are judging patents really don't have the background that the scientist has. On the other hand, we could say that NIH funding tends to mediocrity. Why? The study sections are populated by average scientists. There are very few very creative and gifted scientists that are responsible for the big ideas. Some of them serve on study sections, but if you do it a few times, you're not going to do it again because it takes a whole month out of your productive lifetime to be on one of those study sections. You're asking average people to recognize brilliant new ideas, and it doesn't always happen.

Roger made a selective comparison that parallels this account.[22] When I asked him whether conflicts emerge as a result of ties between universities and industry, he also brought up the NIH:

> The only conflict that I think would be a concern is if industry biases the selection of research problems. If it creates somehow incentives that distort what it is that the university would be doing from some sort of first principles. Having said that, however, I want to say that I think it's a negligible concern relative to NIH. So if you ask me "What's the biggest danger to science?" I consider NIH the biggest danger to science or at least biological science, because it distorts the whole field toward biomedical science, which is a very narrow piece of the biological universe.

The credibility of the commercialists' comparisons is not in question here; what's of interest is not with veracity of the comparison but the necessity. Indeed, that the view they offer may not be limited to commercialists illustrates how selective comparison operates. When challenged, the commercialist is able to point to a *central* aspect of academic work and comparatively frame commercialization as superior to the object of comparison. The result of social comparison is thus a self-justifying mandate for commercial autonomy.

Professionalism

Even if social comparisons are effective, the technique does not alleviate commercialization's own set of problems, namely, ethical issues. When talking about their commercialist activities, commercialists commonly invoke the notion of professionalism—in particular, a professional code of conduct—which they suggest uniquely enables them to deal with the ethical aspects of commercialization in an appropriate manner. To be sure, a formalized ethical code does not actually exist, and thus invoking professionalism is a purely rhetorical strategy, but invocations of professionalism were prevalent among the commercialists I interviewed when they discussed motivations for commercialization, conflicts of interest, and the appropriate balance between research and commercialization in universities. For example, when discussing her views of the "right reason" for engaging in commercialization, an assistant professor of bioengineering in her thirties told me that "it sounds self-serving . . . but we have a little bit more altruistic attitude than the general population in the United States, I believe."[23] "You can trust us" the comment suggests.

This technique seeks to negate the stigma associated with the pursuit of profit by appealing to a general ethical code, even if no such code exists. In contrast to the view that science as a profession is altruistic, some commercialists I interviewed suggested that they possessed a unique moral license that enables them to act appropriately in morally precarious situations. Charles told me:[24]

> There's all kinds of hazards and pitfalls that one can fall into. I'm lucky because I have a strong internal compass that keeps me from getting in trouble on these kinds of issues. I'm basically "science first." When I'm working in my university environment in my university lab, then my job is to do the most interesting and important science as effectively as possible that I can think of doing. And I don't actually care whether it's going to [get] money for me or not.

His comment acknowledges an inherent potential for unethical behavior but neutralizes it by emphasizing, through attribution to luck, the possession of a uniquely attuned moral awareness. Notice that, in Charles's case, this neutralizing is achieved alongside recalibrating the significance of the role played by money (earlier he had said that learning to value money was important in starting a company), such that money has little or no relevance to his actions in the specific context of the university.

Whereas Charles's comment emphasizes individual possession of such moral awareness as unique within the scientific community, other scientists frame the possession of the unique code of professional conduct as institutionally specific. A professor of chemical engineering with more than seventy patents evoked this mode of professionalism when I asked him if universities are getting carried away with the emphasis on commercialization.[25] He told me:

> You can easily say that there are universities that are in the second tier, who are getting too close to being engaged in more contemporary issues to be able to

> support themselves. I think there's a lot of concern in biomedical research communities, now that universities are so engaged in translation for their own economic benefit . . . They may not have as long a view as they need to have, but I think the most valuable universities in the country, and I think I'm sitting in one, and I think Stanford's another, are playing an incredibly valuable role to society. We are criticized for being too close to the private sector. That's fine.

According to this comment, broad moral license is derived from being in an elite university. This implies that scientists at other institutions engage in commercialism for the wrong reason (material gain) and thus abide by an inappropriate ethical code. Relative to other techniques that neutralize or positively imbue work content, professionalism enables a positive work identity by ennobling scientists themselves. The accounts cited here point to moral hazards that individuals and institutions risk encountering and the necessity of recalibrating the role played by financial incentives. By presenting oneself or one's institution as among the few who abide by a moral code that helps them evade such moral pitfalls, commercialists construct themselves and their institutions in a positive light.

Reframing, role distancing, selective comparisons, and professionalism are techniques that resolve the identity instability created by adaptation of the commercial role, allowing commercialists to present a positive self-image despite the pervasive stigma associated with their activities. Internally, these techniques unbind commercialists from the influence of traditional norms, allowing them to persist in the commercialist role. Commercialists, however, are not the only scientists for whom the morally questionable nature of commercialization requires identity work. The emergence of the commercialist role, and potential association with it, also influences elite traditionalists' sense of self-identity because these scientists are situated in commercially intensive environments. I thus turn to consideration of the implications of this stigma for identity construction among traditionalists.

Traditionalist Techniques of Disidentification: Affirmation of Purity

Professional identity is influenced not only by what elite traditionalists *do* as scientists (traditionalist scientific work) but also by what they *do not do* (commercialization). Regarding the latter, identity construction entails "never identities," which are based on exclusion and actions that are possible but not enacted.[26] My application of this concept is unique in the study of science and professions, although the study of this mode of identity work is fundamental to social psychology. In his study of the negative cult, a group that mandates obligations by prescribing abstinences rather than ways of participating, sociologist Emile Durkheim showed that group identity is derived from avoidance of actions that may be tempting (much as financial gain may be) but must be resisted.[27] Acts that individuals do not engage in are central to determining identity. The quintessential "never identity" is the virgin, whose identity as such is predicated on the irreversible consequences for identity of succumbing to the act. Just as the social identity of the virgin is predicated on sexual abstinence, the professional identity of the traditionalist is derived, in part, from commercial abstinence.

In the American academic profession, the notion of a "never identity" is a decidedly modern feature, brought forth by the emergence of the new reward system. For the traditionalist, the introduction of the commercial role, which embodies values and behaviors that are inconsistent with what is traditionally prized, produces a basis for identifying as *not a commercialist*. Traditionalist identity construction is thus based not only on identification with the dimensions of work considered in chapter 1 (what they *do*) but also on disidentification with the commercial role. Traditionalists perceive that the "dirty" nature of the commercial role contradicts valued aspects of the self, and thus they engage in what I refer to as techniques of disidentification, to affirm their purity. Such techniques represent how traditionalists define who they are, based on who they are not. I consider four means by which the process of disidentification occurs: social comparison, disavowal, ritual identification, and retreatism.

Social Comparison

Like commercialists, traditionalists use social comparison in response to the identity challenges commercialization presents, but the technique they use is simply the inverse of the process employed by commercialists. That is, traditionalists disparage referents of commercial culture to portray traditionalism as superior to commercialism. The following account, representative of social comparison among traditionalists, illustrates this. I was discussing industrially sponsored research at universities with a professor of chemistry in his sixties,[28] who stated:

> Directing resources at a defined project is a waste of time, because in order to define something before it's appropriate to define it, you end up working within the incorrect and incomplete knowledge base. The best bang for the buck is

> developing new ways of thinking about things, new methods for doing fundamental tasks, and training people to use their brainpower rather than to harness it to accomplish the idea of a stupid person.

Eager to know more, I asked what he meant by that. He continued:

> Managers can't possibly know very much about science. If a manager, based on his outdated and incomplete education, sets a goal and is too restrictive in his defining of the objectives, you have a person who has tremendous capabilities just dying. When I was an assistant professor, I visited [a high-tech firm] in California, and there was this huge room filled with recent PhDs in carrels, and they were all dying. There was nothing for them to do that was worthy of their PhD. They were prisoners. And after three or four years, they would not be capable of doing anything useful ever again. We train people to define their own problems, not accept problems, and to solve them and to go where the problem leads them.

The professor hones a self-definition while juxtaposing the differences between traditionalist and commercialist science in colorful terms. Traditionalism is presented in a favorable light to indicate. Traditionalists are described as autonomous, while commercialists who accept industrial funding are depicted as subject to external control. The purity of the traditionalist identity is thus affirmed by contrasting it with a "dirty" comparative referent.

The primary distinction between social comparisons made by commercialists and traditionalists is what provokes this mode of identity work. In contrast to commercialists, who made social comparisons when I asked potentially threatening questions about commercial activities, traditionalists engaged in social comparison without provocation. This is indicative of the embattled orientation that traditionalists at elite universities exhibit. They perceive that the value placed on commercialization by their universities and funding agencies threatens their ability to pursue fundamental research.

Disavowal

Another mode of disidentification exhibited by traditionalists I interviewed is disavowal, or disclaiming connection with or responsibility for commercialization. Through disavowal, traditionalists affirm their identity by rejecting commercial incentives as having no binding force on their behavior. Given the continual presence of commercial incentives in their environment, disavowal is a sustained form of identity work in which the purity of one's identity as a traditionalist is reflected in continued rejection of commercial temptation.

The manifestation of such identity work is observed in these scientists' accounts of encounters with commercial incentives, such as consideration of career paths associated with commercialized science, for instance, times when they were approached by commercial or applied organizations offering jobs or research opportunities. I spoke with a professor of chemistry in his early sixties who, when describing his postdoctoral career path, told me

> The first job that I was offered was to work at [a research institute]. I didn't seek that job and I didn't take it, but it was to develop a basic trick that I had used in [my thesis] as a way of developing [an application]. I was not remotely tempted.

The scientist affirms his professional purity in his claim that his research would never be assimilated in an application, even when incentivized by a job offer.

A more telling perspective from which to examine disavowal is the ways traditionalists respond when asked to hypothetically consider circumstances in which they could envision commercializing their work. I asked traditionalists this in interviews to understand conditions of acceptability through having them describe contexts in which they could see commercialization in a more favorable light.[29] That is, the question draws attention to specific incentives that would motivate one to succumb to the temptation to commercialize one's work. The modal pattern in the responses, however, was nearly invariable rejection, succinctly captured in responses such as "It's not for me. I think others will do that,"[30] and "No, not for me personally."[31]

"Not for me" expresses a basic disassociation with the commercialist identity, reflecting the desire among traditionalists to separate themselves from commercialization. This general pattern of rejection can be subdivided into two categories. Some traditionalists responded by repudiating financial incentives, while others contrasted the originality of commercialist and traditionalist contributions.

As an example of the first, I spoke with a traditionalist professor of bioengineering in his sixties about possible circumstances in which he would consider patenting his research.[32] He had been describing a colleague in his department who had accrued $20 million in patent royalties when I told him I wanted to discuss possible motivations of commercialization. He replied:

> I told you my motivation wasn't money. Having a company or even being the chief technology officer is something many of us don't want. It's a big distraction from what you have to do.

> *You were never motivated to attempt to commercialize any of your research?*
> We did some interesting things that actually haven't come to market yet after twenty, thirty years. We made the first portable device that could monitor your [physiological process], a computer-based device. The motivation at that time was and now still would be "let's write a paper about this," so we disclosed all of the technology in publications rather than worry about "let's start a company."

Thus through disassociating themselves from the influence of commercial motives, traditionalists affirm their adherence to the traditional incentive of recognition through publication.

A second pattern of rejection calls into question the validity of the most basic commercial act: patenting. Some traditionalists I interviewed were dismissive of the novelty of discoveries that commercialists patent. For example, an associate professor of chemical engineering in his early fifties was describing to me how unlikely it would be that a patent could result from his work;[33] he paused and said:

> One of the things I have a little trouble with is how different does something have to be to be novel and patentable? I think a lot of the things, some fraction of the stuff that's being patented, are ridiculous, and I can't imagine how the patents are going to stand up.

In this example, disavowal of commercialization is similar to viewing incremental publishing as dirty work. The critique rests upon disparagement of progress—basic or applied—that is barely discernible. Such identity work entails trivializing key referents of commercial behavior, such as the patent, in ways that construct incentives as unpersuasive. The incentive to patent, in this respect, is nonbinding on traditionalist behavior because it is viewed as departing from the traditionalist perception of originality. Such referents of commercial behavior challenge the traditionalist sense of self, and thus their deflection serves to frame them as not associated with the traditionalist identity.

Disavowal affirms the traditionalist identity by severing any perceived tie to behavior associated with commercial incentives. Traditionalists deny responsibility for and association with such behavior by framing commercialization as a practice that others engage in but for them is the basis of a "never identity."

Ritual Identification

Traditionalists may disavow and disparage the commercial role, but the copresence of the two reward systems creates an unavoidable predicament for them.

Referents to the commercial reward system are pervasive. Although many of these referents, such as patents or industrial sponsorships, are easily rejected, others related to the goal of commercialization are less easily avoided due to the structure of funding. When scientists apply for funding, most agencies require them to articulate extrinsic, ancillary goals of research beyond the creation of knowledge. The National Science Foundation, for example, requires that funding proposals discuss the broader impacts of their work which may occur when "projects are applied to other fields of science and technology to create startup companies . . . [and] to improve commercial technology."[34] Other funding agencies require similar criteria when evaluating proposals for funding. Thus, for traditionalists, the identity threat is that they must participate in aspects of the commercial role even though they do not identify with it.

Consider how one professor of chemistry I interviewed at a public university interprets the situation.[35] He was telling me his view of the impact of commercial rewards on funding in science when he stated:

> I have had uniform funding from NSF my entire career. The last few years, they have really upped the broader impact [section] of your proposals. You now have to explain the broader impact of your research very carefully, and many view that as a corruption of the NSF process. I do, actually.
> *In what way is it a corruption?*
> It forces you to commit minor fraud. Somebody who's trying to get greater precision on the bond length of a carbon oxygen bond—that doesn't have societal impacts. And if you say it does, you're basically lying. You're being pushed into the situation where you say "Well, somehow this knowledge will translate into a better drug for cancer." Well, bullshit. You know it doesn't. It's a puppet thing. We're forced to be a puppet in that sense. Different kinds of research have so much more opportunity for this kind of stuff than other areas. Are you going to suppress those areas where you can't make a broader impact statement honestly? I wrestled with this. Something you do that does not have any direct impact on citizens of the United States doesn't mean it's not important to research. That's the part that bothers you. Should we all be working on making better bicycle tires?

The imageries of fraud and puppetry he painted reflect his rejection of the commercialist imperative for societal impact. But the implications of such renunciation entail potential loss of funds for research. This paradox leads traditionalists to engage in what may be called ritual identification, in which the traditionalist participates in, but privately rejects, this minor but nevertheless important commercial act.[36] In other words, traditionalists, absent an abundance of funding for fundamental research, must embrace certain behaviors of the commercial role, but they do so in an automaton-like fashion.

Ritual identification has two related characteristics. The first is that it is perfunctory. An associate professor of chemistry in his thirties,[37] for example, while describing referents to applied science in his papers, told me, "I have the same stock six sentences that I write at the beginning of my papers and then just go on and do my thing." The overtures to commercial goals that he must make are superficially carried out with a minimum amount of reflection.

The second characteristic of ritual identification is that traditionalists do not believe in the ends they perfunctorily express. I spoke with a professor of biology in his late sixties,[38] for example, who at the start the interview told me, "I have been privileged to be able to work on exactly what I want to for my whole life, and that's got to be rare." When I asked him to say more about why he sees his autonomy as unique, he began to critique what he called the "medicalization of science" at NIH, which he said distorts science because "the raw curiosity that drives people gets you farther in understanding basic principles" than "a short turnaround for medically valuable science." As he spoke about applying for funding in this context, he told me:

> I always write, you know, we're looking at sexual reproduction and we will find this pathway through which social influences change the brain. So I can, with some confidence, say girls who grow up in a house without fathers menstruate at a different time by a delta of one year than those who grow up with fathers. That's clearly a social impact on the reproductive cycle. It's not a lie to say we may be able to find out how this is working by finding the pathway, but do I believe I'll find it? No. Do I believe that if we did find it, we'd understand who [should] remedy that? No . . . It's not my thing.

Traditionalists do not perceive ritual identification as deceitful, but they also do not believe that they will achieve the commercial or applied goals they are asked to describe in research proposals. In traditionalists' accounts laden with apathy and relative indifference on this topic, the general sentiment regarding obligatory participation in commercially oriented acts is one of "going through the motions" to get to what is truly valued—funding. This mode of identity management thus requires only a momentary departure from the desired professional identity traditionalists see as their authentic self.

Some traditionalists, however, infuse ritual identification with antagonism. Some of the traditionalists I interviewed were extremely hostile about the "broader impacts" criterion of the National Science Foundation. One associate professor of chemistry was visibly annoyed as our conversation shifted toward discussion of the impact of commercial rewards on funding for science.[39] Like the biologist quoted above, he stated, "I've been given the luxury in my life to pursue science. It's like 'Thank you, world,'" as if he had been given a gift. It was when I followed up by asking if someone with that luxury should be concerned about the possible applications of their research that he winced and responded:

> Fifty years ago, forty years ago, it was pure science. "We think that this is an interesting problem that we'd like to think about." We can't do that anymore. If we go to NSF or to NIH, we have to have an application in mind. Even though it's a pure science pursuit, it's like "We want to figure out a way to tether down dyes onto a surface so we can make dye-sensitized solar cells in a better way." But what's the pure science for me? I'm not really going to make a dye-sensitized solar cell or really try to optimize it. So you write a proposal around dye-sensitized solar cells and say, "We think that a new way of attaching dyes would be something good." Am I really going to spend all my time trying to optimize dyes on solar cells? Don't think so!
>
> *What are the consequences for science of these broader impact statements at NSF?*
>
> I don't even want to talk about NSF. You can talk about any other agency. NSF is just a basket case in my view.
>
> *Why is that?*
>
> I don't want to talk about it. I'll just stay off the record about that.
>
> *You are off the record, period. This is an anonymous interview.*
>
> It's just a basket case. It's not the National Science Foundation anymore, it's all about society building. Okay? Let's take it back to where it's supposed to be, National *Science* [with emphasis] Foundation. If they want to change their name, that's fine. It's not outreach. We already do outreach as it is. We're already over committed as it is. Let's give scientists some money so they can pursue basic science.

Ritual identification is a form of identity work that permits traditionalists to obtain funding: they feign interest in commercialization for official purposes but renounce commercialization otherwise. It is, in effect, a "fact of life" for traditionalists that they must tie their work to extrinsic goals regarded as secondary to the creation of knowledge. For some scientists, this is changing the identity of the academic profession from a community in pursuit of truth to a community engaged in tasks perceived as unrelated to science, such as commercialization and societal outreach. We see in such accounts an embattled group of scientists, who despite their successes seem threatened by the changing conditions of science.

Retreatism

Other traditionalist responses to commercialization suggest the fourth form of identity work: retreatism. In contrast to ritual identification, in which the commercialist identity is rejected by going through the motions, this mode of response entails abstinence and withdrawal. Entertaining commercial incentives is considered "selling out." According to this view, the most honorable stance is to reject commercial means and ends altogether. The purity of one's identity is maintained by reaching the end of one's career free of taint.

I observed this mode of identity work only among traditionalists in late phases of their careers. Exemplifying this view is Lloyd, the confident professor of chemistry we heard from in chapter 2.[40] Lloyd and I were discussing his view of the shift in funding toward technologies and applications when he began to categorize scientists by their motivations for seeking funding. In doing so he provided insight into traditionalists' construction of a "never identity":

> There are a lot of scientists out there and they have nothing to do, they have bad students, and out of desperation, they think, "I'm a clever guy, let's see if I can do something commercial, because there's no way that I'm going to be able to do the kind of fundamental research that I was trained to do as a graduate student." When you look around, there are a few first-rate places where most of the faculty are able to remain intellectually active and doing what they want to do for most of their career. I think that the vast majority of universities in the United States have situations where people cannot keep their research going for more than about ten years, and that leaves twenty to twenty-five years of either becoming bitter and just basically not doing anything or trying to do something that's challenging. Technology is easier to succeed at than basic science simply because all you need is a small thing and then you can work at developing it.

For Lloyd and other traditionalists, the "never identity" is the scientist who engages in commercial research because he is no longer able to acquire funding for basic research. The affirmation of purity for such traditionalists in late stages of their career is retreat: use up one's funding and retire. This is a surprising stance to find at elite universities, where it is not unusual for scientists to continue their work even after taking an emeritus status. As Lloyd explains:

> I have a fairly pessimistic view from my own experience, and you know it makes me . . . I'm not looking forward to retiring but this is one of the reasons why I think that I might like to retire—that I see the kinds of science that I love and am good at being choked off. So I'm pretty pessimistic, and at some point, it will be the thing that leads me to say, okay, I'm turning off the lights in my lab.

Lloyd is an extremely successful scientist, but rather than adapt to a new system of funding criteria, it is important to him to retire with the purist identity he has maintained over the years. In his view, this means success because he is not sacrificing his standards of success and thus, his sense of self.

Traditionalists who engaged in retreatism are among the most successful scientists among the traditionalists I interviewed, yet they are approaching the end of their careers embattled and critical of the changing nature of science. A professor of biology,[41] a member of the National Academy of Sciences, told me that he has "a tiny little dribble left of money" as we discussed funding and the push for applied research. "So what I'm saying is: it's a stupid system the way it's set up. We have to maintain a core of people that aren't motivated toward curing a disease but are just motivated toward understanding what we don't know." Traditionalists engaged in retreatism neither embrace a system they disdain nor engage in ritual identification, actions that could make possible their sustained participation in the research role. Instead, they retreat, pessimistic about the system they renounce but satisfied with their individual career achievements.

Conclusion

Commercialization has destabilized the professional identity of elite academic scientists, challenging how they see themselves and the presentation of self they enact to others. The identity challenges that commercial culture presents to elite scientists are not always dramatic or drastic, but they are pervasive. To manage this challenge, commercialists and traditionalists engage in modes of identity work that project distinct professional identities. Individually, this challenge is consequential because when the views of individual scientists are questioned, those scientists may experience frustration or shame. Commercialists may be viewed as immoral or unethical if they do not learn to project a positive commercial identity. Traditionalists may experience frustration with the difficulty of enacting the professional identity they esteem in environments where commercialization is pervasive, leading some to engage in certain tasks in a perfunctory manner or to retreat from the scientific role altogether. Collectively, the commercial identity challenge is consequential for science because the distinctive identities enacted by commercialists and traditionalists depict a contested collective identity in elite academic science. While that collective identity has never been static or homogenous, the severance of identity within elite universities around the material interests of scientists implies tension and conflict, as traditionalists appear embattled and commercialists appear empowered.

Commercialism, Rationalization, and Fragmentation in Science

According to sociologists Jonathan and Stephen Cole, apart from intellectual controversy, science is an anomaly in society due to its lack of social conflict: "There is very little basic questioning of the legitimacy of the social structure of science by any identifiable group or stratum. There are sporadic cries that rewards are unfairly meted out to individuals, but no organized movement of protest exists against the structure of the reward system or the basis on which the performance of scientists is judged. This relative lack of social conflict is a sociological anomaly."[1] From the foregoing analysis we may conclude that science is no longer as anomalous as the Coles propose. The evidence suggests that nothing less than a transformational shift in the social structure of science has occurred as commercialism has become a fixed presence in academic science.

What are the consequences of a commercially oriented reward system for science, for universities, and for the academic profession? For science, it means access to new technologies, new economic opportunities, and new attempts to solve persistent societal problems. To the extent that scientists and universities are successful in solving such problems, public and political support for science could presumably increase. For universities, this could lead to a stronger resource base to support the pursuit and dissemination of knowledge. For the academic scientists exposed to the mandate for commercialization in their day-to-day work lives, however, the new reward system holds quite different consequences. Departing from the tendencies of scholars to study reward systems in isolation and to underemphasize the subjective careers of commercialists and traditionalists, I offer a new perspective on how we should think about the implications of commercialization for the social organization of higher education. Two overarching conclusions can be drawn from this book: the emergence of commercialism as a new ideology in science and a broader pattern of intraprofessional conflict that destabilizes the role of universities in society.

The Ideology of Commercialism

The increased influence of capitalism within universities has changed the rules and rewards for how science should be done. The new institutional goal of science is to create technologies that have a concrete societal impact. The institutionalization of this goal is the product of pressures in the organizational field of higher education that include policy changes that incentivize the pursuit of intellectual property, federal funding agendas that require proposals to link research to technological and societal outcomes, and structural changes that link universities into increasingly tight relationships with industry. To date researchers have extensively documented the implications of these structural changes for universities and their environments, largely without studying how scientists themselves perceive the new commercially oriented reward system. The irony of this asymmetry of information is that commercialization begins with scientists and is heavily shaped by the professional ideologies they embrace. Having focused on the cultural dimensions of the academic profession—specifically professional ideologies that articulate beliefs about how research should operate and the status systems associated with espoused norms—we are now in a position to summarize the defining characteristics of the ideology of commercialism that is associated with the institutional goal of science.

Most concisely, commercialism is the pursuit of status through societal impact. Commercialism is a type of substantive rationality, or means-end calculation, made in reference to what sociologist Stephen Kalberg calls a "value postulate," that is, an individual's or group's preference for certain ultimate values or actions. For commercialists, the value postulate is societal impact, as seen in the fact that this achievement constitutes the essence of eminence and visibility, and therefore power, in science. Substantive rationality entails a "valid canon," a unique standard against which events are selected, measured, and judged.[2] A value postulate such as societal impact implies clusters of values that organize action. Commercialists' orientations to the dimensions of work codified in the main findings of the study are tied to four defining characteristics: control, efficiency, calculability, and predictability. In table 9, I present these characteristics and the dimensions of work from which they are derived. In the remainder of this section, I characterize commercialism by describing each of its constituent elements.

Before turning to that discussion, however, an important theoretic point merits attention. The elements of commercialism are associated with bureaucratic rationality. In sociologist Stephen Kalberg's formulation of Weberian rationality, this type of rationality is formal, that is, a means-end calculation made in refer-

TABLE 9
The Ideology of Commercialism

Dimension of Work	Commercialist Orientation	Element of Commercialism
Professional product	Technology	Control
Work organization	Hierarchical	Efficiency
Reference groups	Industrial	Efficiency
External positions	Corporate	Efficiency
Problem selection	Utility	Calculability
Visibility sought	Societal impact	Calculability
Product aesthetic	Societal benefits / Tangibility	Calculability
Material benefits	Legitimate	Calculability
Vision of scientific growth	Targeted	Predictability

ence to universally applied laws or regulations.[3] But commercialists do not create technologies because the Bayh-Dole Act or university policies say they must. Instead, their actions are organized by the value postulate of societal impact. In short, while I am describing a new dynamic of rationalization in science originating from the academic profession, I am not making an argument about bureaucratization or what sociologist George Ritzer refers to as "McDonaldization."

Control

The purpose of commercialization—the reason why the end product of commercialist labor is technology—is to control uncertainty. Commercialization seeks to control three dimensions of uncertainty: the uncertainty of societal problems, market uncertainty, and professional uncertainty. Commercialists create technologies because they believe resolving various uncertainties in society requires their expertise.

Control of societal problems, as we have seen, is a central theme in commercialists' accounts of the meaning of and motivation for their work. Contributing to the control of societal problems is a core source of status among commercialists, alongside income and external positions. Each product that commercialists create addresses some aspect of uncertainty in society, whether it be tied to health (e.g., HIV therapeutics), the environment (e.g., biofuel), war (e.g., lightweight protective materials), or other spheres where commercialists and funding organizations believe attention is required.[4]

A more tacit dimension of control is tied to the uncertainty of economic markets. Market control was a key factor in the legislation of the Bayh-Dole Act, and it is critical to the corporations that fund university research, to universities, to federal funding agencies, and to coalitions that are connected to university commercialization. Examples of activities driven by concern for market control are the regional economic development strategies that build, within states, coalitions of government, industry, and universities in an attempt to create jobs, business, and profits. The desire to control economic markets *need not* motivate commercialist research, yet it is not easily separated from a commercial ethos, for such control is a reflection of commercial achievement. The imperative for market success was rarely prominent in commercialists' accounts, but it frequently came up in comments about various aspects of their work, including the number of jobs they had created, the level of revenue their research had generated, and the belief that the market constitutes a type of peer review of the quality of their work. Individual and collective commercial interests tend to coincide in the effort to control economic uncertainty.

Commercialism also seeks to attenuate professional uncertainty regarding, in particular, acquisition of resources for doing research, which many scientists recognize as the most endemic uncertainty of a scientific career. Indeed, "money for the lab" figured prominently in commercialists' discussions of the material benefits of commercialization as a motivation for research. The inverse of this dimension of commercialism was observed among traditionalists in the later stages of their careers who, rather than adopting a commercial orientation to acquire funds for research, retreat from the research role altogether. A second way commercialization is tied to the control of professional uncertainty surfaced in commercialists' discussions of societal views of science. When commercialists commented on whether scientists should be concerned with the utility of their research, the notion that scientists should be able to justify their research to taxpayers reflects a general concern among commercialists about how the profession secures and sustains its mandate. Without such a justification, it is implied, the status of the academic profession in society, and thus the stability of a scientific career, is less certain (a perception tied to a broader concern for distrust of scientists in society).

Efficiency

Efficiency, another element of commercialism, is the capacity to use the optimum means to reach a given end. Three aspects of the study findings reflect how commercialists are oriented to efficiency: hierarchical work organization, industrial reference groups, and positions on corporate boards. As we observed in chapter 2, one of the basic ways in which commercialist and traditionalist moral orders of work diverge rests in the organization of work. Commercialists state that commercially oriented research is considered appropriate for postdoctoral scientists but not graduate students, so commercialists organize their research groups hierarchically, which provides a division of labor that optimally addresses both commercial and academic goals. The efficiency of this work organization lies in the presence of a stratum of positions dedicated specifically to commercial goals. This mode of organization can also enhance efficiency for the commercialist scientist himself, for postdoctoral scientists play a role in the training of graduate students, freeing the commercialist to attend to other commitments. The most efficient form of hierarchical work organization is the research institute, which creates what Eliot Friedson refers to as an administrative professional elite, composed of the commercialists themselves, who organize other professional and staff scientists around a general or specific societal problem.[5] With this kind of organization, some commercialists claim, academic scientists are able to solve "big" problems that could otherwise not be solved by individual scientists.

Commercialists' identification with industrial reference groups reflects their belief that the most effective way to address societal problems is by "grabbing the reins" of corporate organizations that possess the most extensive resources for addressing problems (which reflects how these reference groups may also be tied to the element of control). Such resources include extensive financial capital, expensive analytic instrumentation, a reserve of industrial scientists who perform routinized scientific work considered inappropriate for graduate students, and an organizational apparatus for the distribution and marketing of products. This is why commercialists view industrial sponsorship as enhancing, rather than distorting, the scientific endeavor. Similarly, sitting on corporate boards not only serves as a source of status but also reflects the belief among commercialists that industrial reference groups provide a key conduit through which influence on societal problems can be exercised.[6]

Calculability

Calculability, being able to quantify scientific impact, is perhaps the foremost element of commercialism because of its role in the selection of research problems. The key criterion in commercialist problem selection is utility: whether or not the solution to a problem has a beneficial material impact on society. Commercialists view the traditional outcomes of disinterested science, such as contributions to knowledge or preparation of future scientists, as inadequate because they lack an obvious or immediate economic impact that can be materially quantified. According to commercialists, the utility factor can be calculated, first in the form of a technological product or a company and then in the economic and societal impact the solution to a problem actuates materially.

Calculability is intrinsically tied to the commercial ethos because performance is incentivized monetarily by universities. Visibility provides an expression of the quantifiable value of one's work because societal impact is tied up in the universality of one's inventions. The commercialist whose invention annually brings millions of dollars to a department shines bright, whereas the scientist whose patent remains unlicensed at the technology transfer office is commercially obscure. That is, the breadth of one's influence is determined by the market, and therefore the quest for societal impact expresses the value of calculability. Similarly, the legitimacy of material benefits for the achievement of societal impact signifies the value of calculability.

Calculability is also part of what attracts scientists to commercially oriented careers. Commercialists have an affinity for quantifiable societal impact and the tangibility of the fruits of their labor. Tangibility, we can recall, refers to the material form of one's discovery, such as seeing one's product used in the everyday world, knowing that patients have been helped by a therapeutic drug one has developed, or knowing that a device one has invented is part of a household product such as a television or a computer. For commercialists, knowledge alone "goes into the ether" and thus lacks material substance, whereas technological products may be materially quantified.[7]

Predictability

The fourth element of commercialism is predictability. Predictability is expressed in the commercialist belief that the long-term progress of knowledge should be organized around particular targets: societal problems. Simply pursuing the question "why" is insufficient justification for seeking knowledge. Research should have an *end* beyond knowledge: a point where the research should "end up." The advance of knowledge, commercialists believe, should be directed toward the sources of uncertainty in society.

Predictability is also observable in the extent to which commercialists seek to achieve societal impact by mobilizing the resources of corporations through research partnerships and officer and advisory positions. Such arrangements embody the value of predictability because corporations maintain existing interests in specific problems and then allocate resources to solve those problems, which incentivizes scientific research in a particular direction.

Commercialism and Rationalization in the Academic Profession

The ideology of commercialism is embedded in control of societal uncertainties, efficiency in the organization of work and resources for science, calculability of impact, and the predictability of scientific growth. These patterns are consistent with other studies that document how rationalization is manifest in the contemporary research university, primarily in its bureaucratic organizational base.[8] Bureaucratic dynamics of rationalization within universities can be seen as resulting from the historical process of economic rationalization of political life that Berman identifies, in which the substantive rationality for supporting academic science shifted from an emphasis on expanding knowledge and meeting national needs to strengthening innovation and economic productivity.[9]

From Weber's perspective, capitalism and bureaucracies provide the quintessential basis for rationalization processes, but the emergence of elements of formal rationality within the academic profession is noteworthy because rationalization is presumed to conflict with professional organization. In an influential article, sociologists George Ritzer and David Walczak argue that the spread of formal rationality contributes to deprofessionalization, or the erosion of professional control.[10] Using the example of medicine, Ritzer and Walczak show how government regulations and a profit motive in medicine deprofessionalize the medical profession because formally rational values such as rules and regulations about medical delivery erode substantive rationalities such as altruism and autonomy. In turn, they argue, the decline of substantive rationality in medicine leads to deprofessionalization because there are no longer differences between medicine and formally rational bureaucrats and capitalists.

In science, however, formal rationality does not constrain professional control because commercialism is not an end but a means by which scientists pursue status and societal impact (but bureaucratic formal rationality can contribute to the erosion of professional control). And to the extent that commercialism blurs differences between scientists and capitalists, the professional control of elite scientists is not weakened; rather, *it becomes even stronger*, as commercialists are able to maximize social and economic benefits (status, resources for research, royalties) through their market endeavors. Indeed, rationalization empowers commercialists in that they command resources of vast bureaucratic organizations that impact the medicine we take, the way we use energy, and the markets in we which participate as workers. Sociologist Donald Levine points out that rationalization occurs in different spheres of social life and brings out the distinctive values of each sphere according to its inner logic. In the academic profession, the new ideology of achieving status through societal impact leads some scientists to embrace a mode of rationalization that seeks to improve the world by solving societal problems using scientific knowledge.[11] Indeed, professional-ideology-based rationalization takes quite a different form than the bureaucratic logics that some scholars suggest are making universities operate like fast food restaurants. Nevertheless, as we widen our focus to consider the implications of commercialization for both traditionalists as well as commercialists, we observe that professional control is threatened by internal conflict within the academic profession, particularly as it relates to what types of power scientists wield and the differential allocation of rewards for achievement in science.

Intraprofessional Conflict

Commercialism is part of a broader pattern of intraprofessional conflict between commercialists and traditionalists defined by polarization in the norms that scientists espouse, the forms of status they seek, the career paths they pursue, and the identities they construct. Commercialists and traditionalists are not "at war" with each other, but they are divided in ways that depict contested orders of science. The contest centers on ideologies of scientific work: how career paths within higher education should be constructed, what constitutes attainment, and how rewards should be allocated. The presence of competing reward systems severs the stability of moral orders within specific organizational contexts (within departments and universities), thereby fostering conflicting interpretations of careers.[12] The result is that the academic profession is pulled into two different directions, which presents problems for the legitimacy of universities.

This "tug of war" compounds existing fragmentation in the structure of the academic profession, with sources of differentiation wielding both positive and pernicious influences. Burton Clark referred to the professoriate as the "fragmented profession," emphasizing its numerous disciplines, subdisciplines, and specialties within subfields. Disciplinary fragmentation can be an asset because it provides the diverse expertise essential to the development of an enlightened and productive society. But it can be problematic for the production of knowledge when potentially complementary areas of knowledge are excessively walled off from one another or when disciplinary status cultures trump collegiality.[13] The professoriate is also fragmented by institutional type. The vast diversity of colleges and universities (e.g., research-intensive, baccalaureate, two-year colleges) within systems of higher education is an asset because it allows an equally diverse range of societal missions to be fulfilled, such as the production of knowledge, the education of first-generation college students, and preparation for elite roles in society. But this diversity also means that the careers of faculty within the same field at different institutions may bare little if any resemblance to one another, particularly in terms of the rewards, power, and influence obtained by faculty. Perhaps the most powerful source of fragmentation has been the expansion of nontenured and part-time faculty roles.[14] While the use of contingent labor provides universities with organizational and financial flexibility, these positions offer limited rewards and challenging conditions of work.

Commercialism is a cultural fracture that has emerged alongside these longstanding structural sources of fragmentation. Economic and status pursuits can weaken academic community, as the influx of commercial opportunities in higher education has led to distinctively different career scripts. The traditionalist career script is predicated on the moral mandate of science being the discovery and transmission of knowledge. Espousal of communalism, universalism, disinterestedness, and organized skepticism reproduces the institutional logic of science and allocates recognition to scientists who uphold these norms. Power among traditionalists is derived from one's contributions to knowledge—the "units" through which scientists interactively achieve tangible power and exercise authority—and expressed in the caliber of one's institution, one's academic rank, and federal and organizational resources for research. Upholding the institutional mandate through contributions to knowledge yields material power (although such power significantly diminishes beyond the university) and is equally tied to a rich symbolic universe in which power is objectively signified in scientific visibility. Eponymy and immortality, for instance, operate institutionally as a symbolic system that encodes how traditionalists make sense of their role. At the interactive level, this achievement is manifested in an expression that is at once simple to understand and difficult to fulfill: "Change how we think." Embracing this mandate involves an interpretive career modality in which the construction of identity revolves around the affirmation of purity, so as to avoid the sanctions resulting from the perception that one is committed to anything other than the advance of knowledge.

The mandate of commercialism, by contrast, is to control uncertainty through the development of technologies that address societal problems and lead to economic development. This institutional mandate encourages the behavior of commercialists through the allocation of material and symbolic rewards for adherence to the elements of commercialism: control, calculability, efficiency, and predictability. Just as the traditionalist is unable to receive recognition without adherence to communalism, the commercialist is unable to receive financial rewards if he or she does not embrace calculability, or the emphasis on quantifiable material impacts on society. Power is achieved through creating commercial intellectual property in the form of patents—the essence of the commercialist role—and it is exercised wielding authoritative influence as advisors, directors, officers, and founders of corporate firms. Career scripts among commercialists, organized around visibility through societal impact, entail a rich interpretive scheme of techniques that legitimate working outside of the organizing knowledge of the profession.

These distinctive traditionalist and commercialist career scripts are not simply alternative paths taken in science. They are tied to the restructuring of the institution of science, to status within it, and to the bases of stratification and inequality. In fact, the significance of traditional status attainment in science, much like the professional identity to which it corresponds, has been destabilized by the presence of a competing alternative. Commercialist and traditionalist attainment could coexist within science without disrupting power relations, but social systems tend to organize around hierarchical rather than egalitarian structural forms. We must therefore consider how commercialization may be related to new forms of stratification in the academic profession.

Commercial Attainment

Commercial attainment as the basis of scientific status is of importance to social inequality among scientists, for status yields influence. Status provides an evaluative hierarchy among social groups and among individuals.[15] As we have observed, competing career scripts are tied to status beliefs held among commercialists and traditionalists, who associate greater social esteem and value with differing definitions of what science is and what it should do. To the extent that it is associated with variance in conditions of work, definitions of worth, and the allocation of rewards, commercial attainment threatens traditional status in science in its potential function as an axis along which social relations are organized.

The presence of commercial attainment as a new form of status in science suggests a fertile context for *competing* status hierarchies. A status belief forms when those in the favored social category and those in the disfavored category agree, as a matter of social reality, that group members of the favored category are accorded greater respect than those in the disfavored group.[16] The question therefore emerges: Do traditionalists concede that commercialists are seen, within the profession and society, as better than traditionalists? Without question, although they do not personally endorse such a view, traditionalists view the major scientific organizations, universities, and funding agencies as favoring commercialists. Traditionalists similarly believe that the public lacks appreciation for the value of basic science.

The ascent of commercial attainment in the status hierarchy of science suggests that a new basis of anomie may be under way in the academic profession. Durkheim introduced the concept of anomie to describe a condition of deregulation in society, where the norms that usually govern expectations for behavior are not present.[17] In the academic profession, anomie arises through circumstances in which scientists lack opportunities to achieve recognition, where cultural conditions that foster commitment to professional norms are low, and where organizational expectations are inconsistent with the activities valued most by individual faculty members.[18] Commercialism contributes to anomie in that the new mode of status attainment has, among some segments of the profession, weakened commitment to traditional norms. Traditionalists—particularly those at elite institutions—still possess opportunities to achieve recognition, but commercialism has an elevated status in the reward system because universities, funding agencies, and many elite scientists place a higher value on commercial attainment.

Indeed, traditionalists see commercialists as holding higher status. One such sign is the pervasive concern among traditionalists regarding unequal conditions of work. They comment that their commercialist colleagues' commitment to corporate firms or commercially tied or funded research institutes shifts ancillary role "burdens," such as departmental obligations and teaching, to traditionalists. Or they note that commercialists can "write their own check." Both patterns suggest the potential for unequal conditions of work that vary by workload and nature of task. It also appears that commercialists, independent of their royalties, earn higher salaries than traditionalists. With royalties, which are frequently higher than a nine-month academic salary, included, commercialists very clearly earn more than traditionalists. Such distinctions reflect differing levels of status, which potentially impacts job satisfaction among traditionalists, who may view themselves as less valued by their universities. Finally, traditionalists tend to view themselves as threatened by the emphasis placed on commercialization, by both universities and funding agencies. All of this evidence suggests the ascent of commercial attainment in the status hierarchy and a new basis for anomie among traditionalists. This pattern is reflected in much of the data from traditionalist interviews and is seen particularly in the techniques traditionalists employ to disassociate themselves from commercialist referents, retreatism being the most extreme. Indeed, many traditionalists I interviewed seem to feel, despite their successes, embattled.[19]

Contested orders of science and competing status hierarchies paint a picture of a fractured profession in which one segment of the profession seeks an exclusively professional mode of control over the definition and coordination of work, while another views market and bureaucratic forms of control as enhancing rather than constraining their power. Academe is thus characterized by what Abbott calls an internal jurisdictional conflict, which may lead to a less effective claim to jurisdiction.[20] Indeed, at present, the academic profession has already ceded some of its power over the direction of the production of knowledge. It is unquestionable that industrial sponsorship, commercially oriented federal sponsorship, and lower levels of consensus within fields surrounding the selection of research problems are undermining the ability of the profession to exclusively designate the direction of knowledge. The future of the profession's jurisdictional claim will depend in part on the extent to which it allows further corporate patronage onto its turf.

This is, however, a two-sided issue. Successful commercialists do not interpret their alliances with industry as an encroachment on their professional control. In fact, they believe such alignments enhance the reach and quality of their scientific work. But if industrial support becomes a growing form of patronage in science such that it competes with federal funding, professional control will weaken. In that scenario, scientists who otherwise reject commercially oriented work would wield less influence on the definition of problems that should be studied and the manner in which such work is carried out. This development would evoke a second factor that influences the efficacy of academe's claim to jurisdictional control, one over which it wields relatively little influence: the economic outlook for higher education. A "checklist" of financial uncertainties that pervade higher education in times of economic uncertainty includes falling endowment values, lower state appropriations, lower enrollment revenues, and federal budgetary pressures.[21] The continuation of this economic circumstance acts to the detriment of both commercialists and traditionalists because both federal funding agencies and research-intensive corporations are subject to economic downturns. But traditionalists are likely to suffer more, if only because commercialists have alternative (industrial) sources of funding they would accept with little hesitation. The perpetuation of an economic downturn could thus destabilize the jurisdictional control of the academic profession to the extent that the opportunity to perform research is increasingly tied to whether one can attain industrial funding. In that instance, the primary avenue of encroachment into jurisdictional control would be tied to the selection of a problem, which precedes technical control of work.

There are two ways of thinking about what this scenario means for society. One is that universities will continue to be much more engaged in the production of technologies that are good for human well-being and the environment and that provide new jobs, expand new markets, and contribute to the economic health of the nation. But corporate influence will be imperfectly aligned with the public good. Oil corporations could favor funding research on technologies for oil exploration rather than technologies for containing oil leaks, for example. Agricultural biotechnology corporations could sponsor research that creates technologies helpful for food production but that helps major farming corporations while hurting independent farmers. The point is that even when contributing to societal welfare, corporate-sponsorship of university research will reflect interests that coincide with corporate sponsors' bottom lines, which will inherently favor specific technologies and markets over others. Commercialists have the opportunity to direct that influence in positive ways.

Policy Implications and Recommendations for Institutional Action

The study of commercialization is tied to a policy framework because the institutionalization of commercialization in academe was originally catalyzed by federal policy. As Slaughter and Rhoades point out, while the Bayh-Dole Act and other federal patent laws and policies do not dictate commercialization, they create opportunities for it.[22] Universities and the scientists who constitute them are thus not "corporatized" or acted upon in this sense. Rather, scientists in the academic community actively participate in commercialist activities with the mandate of judicial law and federal and state policies. Thus, the implications of my findings for policy should be considered not in terms of the relative advantages or disadvantages of a federal technology transfer policy for innovation or economic development but instead in terms of general research funding policies that support a technology transfer regime and organizational policies related to the impact of commercialization on work within universities.

Before turning to funding and organizational policies, however, a consideration of the efficacy of the Bayh-Dole Act merits our attention, given its thematic relevance to the issues in this study and the viewpoints of the scientists. If the goal of the Bayh-Dole Act is to promote discoveries that bear on societal problems, how effective is this policy? One way to address this question is to consider it in the context of a specific area of concern, such as human health—a useful choice given the volume of commercially oriented research that addresses it. Two illustrative cases call into question the effectiveness of the role of patenting scientific research in the creation of therapies for or solutions to human illness. The first comes from a survey, conducted by the *British Medical Journal*, that asked British physicians to rate the most important medical and pharmaceutical "milestones" in history.[23] Of the resulting top fifteen discoveries, only two (the birth control pill and chlorpromazine) were patented or resulted from a previous patent. The second case, a list of top-selling pharmaceuticals worldwide published by *Chemical and Engineering News* magazine, makes a case for patenting but not a powerful one.[24] One analysis of the forty-six top pharmaceuticals finds that, of these drugs, twenty lack a patent or are not based on existing patents, four were discovered in the process of curiosity-driven research, and two were discovered in an American university labs prior to the Bayh-Dole Act.[25] In short, the analysis suggests that half of the top-selling pharmaceuticals in the world do not owe their influence to patents. These two cases reflect the position among traditionalists' that commercial incentives are not necessary for scientific research to produce discoveries of societal relevance. And if we take a disciplinary perspective and focus on a discipline such as physics, we see that basic research with a long-term perspective has similarly produced some of our largest economic revolutions: quantum theory led to semiconductors and computer chips, while the multibillion-dollar growth industry tied to global positioning systems is sourced in Einstein's work on relativity at the beginning of the twentieth century. It appears, then, that research policy should not emphasize one mode of inquiry to the detriment of the other.

As we have seen, however, federal funding research policies clearly emphasize targeted research. The funded scientist may engage in basic research, but only if the research problem can be tied to a societal problem, such as finding a cure for Alzheimer's disease or training K-12 science and math teachers. The results of this study suggest that policies that have as their objective the treatment of such problems would benefit from giving greater consideration to scientists' interpretations of their work. On one hand, both traditionalists and commercialists consider the solution of societal problems important, and neither group needs a regulatory roadmap to move them in that direction. On the other hand, these two groups of scientists embrace distinctive visions of such a "laissez-faire" science.

Among traditionalists, "free market" science entails solving societal problems through the autonomous pursuit of scientific problems. This view is predicated on the abundance of curiosity-driven scientific discoveries that have had tremendous implications for solving societal problems, such as those referenced above.

This perspective is likewise embedded in the traditionalist reward system. In the view of traditionalists, science needs no external regulation because it already has a system of rewards that operates well. Frivolity in research problem selection rarely, if ever, bears rewards, while working on problems deemed important by past research and by experts in a field—the traditionalist definition of utility—is seen as pushing back the frontiers of science, often with great implications for societal outcomes. The social constraint in this situation, however, is that society is unable to evaluate whether a certain vein of research is frivolous or not. As we saw in a scientists' discussion of the case of Lee Hartwell in chapter 1, whose research into the mutation of yeast proved crucial to the understanding of cancer, seemingly frivolous but scientifically important problems can turn out to be of tremendous value to societal problems.

Among commercialists, "free market" science entails a freedom they already enjoy, the targeted pursuit of societal problems, but free of a programmatic agenda determined by federal funding agencies. As we observed in commercialists' use of social comparison as a mode of identity work, some commercialists disparage federal funding as a greater threat to science than industry is. They view industrial reference groups as enabling the freedom to pursue lines of inquiry they consider important and federal funding agencies as a source of distortion in science.

Thus a point of convergence emerges among traditionalists and commercialists: the view that federal research agendas distort the direction of science. Traditionalists would also characterize industrial influence as a source of distortion, but they are free to reject economic incentives—though only to the extent that they can link their research with commercially relevant targets and ceremonially identify with such goals.

The policy risk produced by federal research programming of societal targets is therefore the suffocation of important lines of inquiry in fields of research that are not clearly relevant commercially yet are potentially important to known and unknown problems, now and in the future. Federal funding agencies need not incentivize particular problems because the opportunities for commercial attainment created by the Bayh-Dole Act already allocate rewards to such problems for the subset of scientists who value such rewards. Were federal funding agencies to attenuate their emphasis on societal targets, commercialist and traditionalist reward systems could potentially coexist with less tension. Otherwise, fundamental research suffers under the strain of both federal and organizational funding systems favoring commercialist research. It is important to encourage scientists to think about how their work affects society, but policies that push this goal too far are likely to undermine the sought-after objective.

A second area of policy informed by this study pertains to how universities organizationally manage commercialization and its consequences. This is critical to society because no matter what types of knowledge they produce, universities are institutions of higher learning. Here the concern centers on the ability of universities to shape two commercially related conflicts of commitment: between commercial involvement and commitment to departmental obligations, and between commercial involvement and doctoral training. Both issues concern how universities weigh commercial activities against the provision of quality graduate and undergraduate education. The university's expressed commitment to undergraduate education is not always reflected in its policies; for instance, the decision to hire adjunct professors and part-time instructors to teach undergraduate courses and thus release commercialists from teaching loads—a policy in which the university benefits from both commercialist royalties and lower instructional costs. This kind of policy is less likely to affect doctoral training. However, comments by traditionalists and, to a lesser extent, commercialists interviewed in the study suggests that the involving nature of commercialist laboratories often results in postdoctoral scientists assuming oversight roles in the training of doctoral students.

Other conflicts of commitment tied to doctoral training include the allocation of commercialist research tasks to doctoral students and secrecy within research groups. Current university policies regarding oversight of doctoral training, frequently explained to me by many commercialists, typically include a written and oral disclosure to a commercialist's research group of his or her commercial interests and the alternate avenues that may be pursued if a student feels such interests are detrimental to his or her training. When asked how effectively such disclosures and other policies concerning external commitments are able to regulate behavior, most scientists replied "not much." Since close oversight of the scientific administration of research groups and time commitments would likely be interpreted as an intrusion into professional autonomy, universities are challenged to construct ways to prevent misconduct in the context of graduate training and to ensure that scientists do not eschew their departmental commitments. A useful framework within which existing policies could be optimized might include initiating committees within departments so that regulation is more localized and peer-based—in contrast to financial conflict-of-interest committees, which are typically university-wide.

Anyone who has studied federal funding for research and state appropriations for higher education recognizes that higher education needs both corporate funding and federal funding and a combined pursuit of basic knowledge of the world around us and technological solutions to societal uncertainties. Given the concentration of expertise and creativity in research universities, these institutions equally possess a moral imperative to tackle society's greatest challenges. But any approach to commercialization that undermines the pursuit of basic knowledge and further contributes to "haves" and "have nots" in the academic profession is short-sighted and detrimental to conditions of work in higher education.

Several groups have important roles to play in creating the basis for an integrated faculty that collectively pursues basic knowledge and societal impact through technological innovation. Professional associations such as the American Association for the Advancement of Science and disciplinary associations should commit resources to communicating the value of basic research and enhancing public understanding of science. Politicians view academic science as an economic engine. Lobbying by professional associations and other constituencies in the scientific community could contribute to a better understanding of the long-term economic gains that come from investment in research that has no clear application or connection to a societal outcome. Politicians would also benefit from a better understanding of the near-term economic impact that results when scientists trained in basic laboratories pursue careers in industry. Such lobbying might help attenuate the existing requirement to connect one's research to broader impacts or preclude more stringent requirements in the future.

University presidents, provosts, and trustees need to recognize that very few faculty innovations become financial blockbusters, and that on average, rates of return to commercialization are low. Over the thirteen-year period between 1996 and 2008, the ten universities with the highest licensing income had rates of return (total licensing income divided by total research expenditures) that ranged from 0.3 percent (the University of Washington and the University of California System) to 4.3 percent (New York University).[26] Similar data analyses conclude that at the average university, only one out of 200 active licensing agreements will generate more than $1 million in royalty income, while all of these agreements would require funding for staff, operations, patent filing, and licensing costs.[27] Beyond merely recognizing the challenges to commercial success, senior administrative leaders can play a key role in how major partnerships with science-based corporations are structured to ensure that any commercial funding arrangements are contingent on substantial contributions to basic research, such as funding lines for doctoral and postdoctoral training in laboratories that lack specific ties to funding corporations. This in turn could improve traditionalists' impressions of institutional ties to corporate science and commercialism in academe.

Deans and department chairs have an important role to play in protecting the well-being of doctoral and postdoctoral scientists to ensure timely completion of training, protection of their own intellectual property rights, and enforcement of conflict-of-interest policies. Rather than simply requiring commercialists to inform scientists in their lab of potential conflicts of interest, deans and department chairs should create a college-wide standing advisory committee composed of commercial and noncommercial faculty who routinely review practices within laboratories. Deans and department chairs should work toward policies that reward commercial pursuits as part of consideration for promotion and tenure. Relative to recent decades, many more faculty enter the professoriate who embrace commercialism. If forced to pursue commercial opportunities on the side, many talented scientists may forgo academic careers or leave before attaining tenure. To provide junior faculty with the time and space necessary for commercialization, and to do so without alienating noncommercial faculty, deans could create equivalent competitive fellowships for commercialists and traditionalists in which junior scientists are relieved of teaching and service commitments for a semester as a reward for major achievements. Importantly, the creation of organizational reward systems for commercial success must avoid creating a "caste system" in which traditionalists cannot attain equivalent conditions of work or in which commercialists can avoid standard departmental and university commitments.

Of course, scientists themselves have an important role to play in promoting a professional culture in which commercialism and traditionalism thrive alongside each other. Communication of the societal value of their work is an equally important task for commercialists, in part to help dispel myths that commercialism is exclusively motivated by the pursuit of one's financial self-interest. But talk is cheap, so commercialists who partner with industry (or through their own startups) could pursue arrangements that require revenue streams be distributed to core departmental basic science missions, such as doctoral and postdoctoral training. Perhaps most importantly, commercialists should not allow their commercial endeavors to undermine their commitment to their graduate students, postdoctoral scientists, departmental colleagues, or teaching and service requirements. For traditionalists, communication of the value of their research and the ways in which it contributes to both commercial and noncommercial outcomes is essential in a funding climate in which the federal government is the only source of support for basic research. Traditionalists should work with commercialist colleagues and university leaders to promote corporate sponsorship of basic research that lacks strings or encroachment on professional autonomy.

My objective in this book has been to develop an understanding of commercialization based on the interpretations of scientists—a unique approach, because most of what we know about the commercialization of academic science is derived from research that has underemphasized or altogether ignored the perspectives of scientists. My findings have expanded the work of previous researchers by offering a view of what commercialization means to both commercialists and traditionalists and by identifying and explaining the dimensions along which these views are at odds. In addition, I have developed lines of inquiry heretofore absent or undeveloped in this literature, including the contours of normative tension, the redefinition of status in science, the cultural antecedents of commercial career trajectories, and the identity work that commercialists and traditionalists enact as a result of a new reward system. The result is the identification of social conflict in science previously unseen.

One of science's greatest strengths, its universality, could also be its Achilles' heel. The fruits of science contain both natural and societal applications, but the quest to extract such fruits by any logic that undermines the advance of knowledge could surely be to the detriment of society by severing the meanings, norms, and identity that make academe an integrated, rather than a fractured, profession. The organization of scientific work should therefore be exclusively wed to neither a traditionalist nor a commercialist approach to science. Science's status in society and what universities do will continue to be shaped by both the relatively old and the relatively new reward systems and by the day-to-day lives of scientists as they seek meaning and success at work.

Interview Protocol

This study is about the reward system in academic science. The questions I would like to discuss concern one's conception of the scientific role and the commercialization of scientific discoveries. Some of the topics I address ask you to make personal judgments about your own career and various professional issues. Your participation in this study is strictly confidential. With your consent to begin, I will record the interview. Interviews are normally recorded to keep track of information accurately. Subsequently, the tape will be destroyed. Your identity and that of your institution will be carefully concealed in any published work. Your participation in this study is important. However, should you at any time wish to stop, you may do so without consequence, and at any time you should feel free to ask me questions concerning the interview or the study. May we begin?

A. Conception of the Scientific Role

1. Among the types of pursuit that university scientists can engage in, what, to you, is the best or most esteemed?
   Probe: Why?

2. To what extent do you perceive your own work to fulfill the ideals that underlie these types of pursuits?

3. Who do you think benefits from what you've done?

4. Which types of pursuit do you shun or regard as trivial?
   Probe: Why?

5. Is commercializing scientific discovery a legitimate way to fulfill the academic role?
   Probe: Why or why not?

6. What distinguishes you from peers in your field who do, or do not, commercialize their work?

B. Motivations of Entrepreneurialism

7. To what extent have you considered or sought to apply your work commercially over the course of your career?
   Probe: Aside from patents or startups connected to your own research, what ties do you maintain with industry?

8a. [*Traditionalists*] Under what circumstances, if any, would you attempt to commercialize your work?

8b. [*Commercialists*] What factors influenced you to commercialize your work?

9. What conceptions of university-based commercial research did you have as a graduate student?
   Probe: In what ways were you exposed to it during this period?

10. In what ways [has/would] commercializing your work allow[ed] you to achieve your goals as a scientist that [aren't/wouldn't be] possible by other means?

C. Norms of Science

11. What should be the relationship, if any, between research and commercialization?
    Probe: Do you believe this relationship gets abused? If so, how?
    Probe: Do you believe universities have gotten "carried away" with this relationship? If yes, How so?

12. Is there a collective understanding among your peers in this department concerning the appropriateness of commercializing one's work?

13. Should universities reward scientists for commercializing their work?

14. Do you see any problems associated with academic researchers accepting money from industry to conduct research?

15. Do you perceive there to be flaws associated with basic science as a mode of inquiry?

16. Should researchers be concerned with the utility of their discoveries?

D. The Operation of Reward Systems

17. Do you perceive costs to the commercialization of research?

18. Do you believe university-based commercial research corrupts science?
    Probe: Are there examples in your own career where commercial involvement interfered with or was incompatible with scientific progress?

19. How do you perceive that the privatization of scientific discoveries by peers in your field influences scientific progress?
    Probe: What would happen to scientific progress in your field if the

majority of academic scientists working in it sought to commercially exploit their work?

20. How can science benefit from increased commercialization of academic research?

21. How important is it to you to make a lot of money?

22. Do you feel that the rewards you have received have been commensurate with the amount of research you have contributed to your field?

23a. [Traditionalists] What do you think would be the most rewarding aspect of commercializing your work?

23b. [Commercialists] What is the most rewarding aspect of commercializing your work?
Probe: Why?
I have one remaining question I would like to address.

24. What have you found to be the most rewarding accomplishment of your career?
Probe: What sets this achievement apart from other accomplishments?
Probe: What was most rewarding about this accomplishment?
Probe: [For entrepreneurial scientists] How does this achievement compare to the success you've attained in your [commercial/non-commercial] endeavors?

Introduction · *Professional Ideologies in Higher Education*

1. Berman 2011.

2. Rapoport 2006.

3. National Science Foundation 2014; Rapoport 2006.

4. Council on Government Relations, 1999; Henderson, Jaffe, and Trajtenberg, 1998; Pressman, 2002.

5. Commercialists have not been excluded entirely from qualitative research. Slaughter and colleagues (Slaughter and Leslie 1997; Slaughter and Rhoades 2004) incorporate interview data that highlight costs, benefits, and manifestations of commercial activity in the academy; Etzkowitz's pioneering work (1983; 1989) in this area focuses on scientists with limited commercial involvement relatively soon after the passage of the Bayh-Dole Act; and Vallas and Kleinman (2008) compare the experiences of life scientists in academe and industry.

6. An important exception to the neglect of scientists who eschew commercialism is a study by sociologists Jason-Owen Smith and Walter Powell that employs deductive logic to create a typology of faculty responses to commercialization. Owen-Smith and Powell cluster faculty into four ideal types: "old school" traditionalist scientists; "new school" scientists; "reluctant entrepreneurs"; and "engaged traditionalists." Focusing on four scientists who represent each of the four ideal types, Owen-Smith and Powell identify patterns of conflict and agreement among faculty responses to commercialization. The Owen-Smith and Powell model nicely highlights the complexities among different groups of life scientists. See Owen-Smith and Powell 2001.

7. For another critical contribution laying the framework of the study of academic capitalism, see Hackett 1990.

8. Geiger and Sá 2009.

9. See, for example, Bercovitz et al. 2008; Siegel, Waldman, and Link 2003; Thursby, Jensen, and Thursby 2001.

10. Slaughter et al. 2014.

11. Kirp 2003.

12. Rhoades 2014, 113–14.

13. Slaughter 2014.
14. Trice 1993, 46.
15. Thompson 1984, 4.
16. Di Gregorio and Shane 2003; Zucker and Darby 2007; Zucker, Darby, and Brewer 1998; Fleming, King, and Juda 2007; Fleming, Lee, and Frenken 2007; Shapira, Youtie, and Arora 2012.
17. Stephan et al. 2007.
18. Di Gregorio and Shane 2003; Etzkowitz 2010.
19. Geiger and Sá 2009, 16.
20. Bok 2003; Krimsky 2003; Slaughter and Rhoades 2004.
21. Merton 1973.
22. Johnson and Hermanowicz 2017.
23. Etzkowitz 1989.
24. Etzkowitz 1998; Colyvas and Powell 2006; Metlay 2006; Stuart and Ding 2006.
25. National Science Board 2004; Stephan et al. 2007.
26. Etzkowitz 1989 and 1998; Owen-Smith and Powell 2001; Colyvas and Powell 2006.
27. Etzkowitz 1989, 14.
28. Etzkowitz 1998, 828.
29. Other research is suggestive of commercial norms but is unable to tap into the constellation of values that compose these new norms. Several quantitative studies that attempt to assess the importance of norms for commercial behavior find that "local norms" are an important determinant of commercial behavior. These studies, however, use broad proxies of norms, such as the number of patents at a university or being situated in a department whose chair is a commercialist. See Bercovitz and Feldman 2008; Louis et al. 1989; Stuart and Ding 2006. One study attempts to model "market values" more directly by measuring scientists' perceptions of the appropriate influence of industry on research, but this approach imposes meaning rather than identifying the core dimensions of how scientists construct norms. See Glenna et al. 2011.
30. Hodson 1999, 294.
31. Durkheim (1915) 1995.
32. Hermanowicz 2009, 25.
33. Merton 1973, 521.
34. Ibid.
35. Abbott 1988, 188.
36. Siegel and Phan 2005; Phan and Siegel 2006; and Siegel, Veugelers, and Wright 2007.
37. Slaughter and Leslie 1997; Slaughter and Rhoades 2004.
38. Azoulay, Ding, and Stuart 2007; Bercovitz and Feldman 2008; Whittington and Smith-Doerr 2005; Stuart and Ding 2006.
39. Göktepe-Hulten and Mahagaonkar 2010; Lam 2011.
40. Stephan 2012.
41. Hughes 1951.

42. Leidner 2006, 451.

43. Clark 2000; Kogan 2000; Henkel 2005. Certainly consulting existed, certainly biotech was on the rise in the 1970s, and engineers celebrated application, but overall patent data clearly illustrate that commercialization was generally avoided. See, for example, Henderson, Jaffee, and Trajtenberg 1998.

44. See Bok 2003; Krimsky 2003; American Association of Medical Colleges 2002; Association of American Universities 2001; U.S. Department of Health and Human Services 2009; Washburn 2005; Skloot 2010; Blumenthal et al. 1996; Bekelman, Li, and Gross 2003; and Grassley 2008.

45. Tajfel 1982; Tajfel and Turner 1979.

46. Lam 2010.

47. Collins 1975.

48. Freidson 1970.

49. Bourdieu 1984.

50. See, for example, Colyvas and Powell 2006. To examine this critique in greater detail, see Aldrich and Ruef 2006 and Fligstein and McAdam 2012.

51. A full description of the study design, method, and research instruments is located at http://www.unr.edu/education/faculty/johnson.

52. The fourth research question related to identity is thus a post hoc research question from the research as it unfolded.

53. Although not all patents become products, patents that do are the most reliable indicator of commercial behavior available.

54. Some perspectives suggest that the mechanisms by which scientists embrace or eschew commercial roles varies across public and private universities, as these institutional types exhibit distinct organizational reward systems and infrastructures for technology transfer. See, for example, Alexander 2001; Cho et al. 2000; Friedman and Silberman 2003. Despite these arguments, I found almost no substantive differences in the views of public and private scientists in my analysis and therefore do not pursue such comparison in the book. This "nonfinding" is likely because the "effect" of elite status outweighs differences in institutional type.

55. This percentage reflects the scientists I either contacted by phone or who contacted me before I could call who agreed to participate in the study. This rate of response is only slightly lower than other interviewed-based studies of scientists, whose response rates were approximately 70 percent. See Gaston 1978; Zuckerman 1977; Hermanowicz 2009.

Chapter 1 · Normative Tension in Commercial Contexts

1. Goffman 1967.

2. Organized skepticism, a fourth norm outlined by Merton, was referred to by traditionalist scientists only minimally and indirectly, mostly in the form of comments that commercialization requires "cheerleading," "marketing," or "pushing what is currently hot" by scientists in ways that counter the notion of detached scrutiny of discoveries. Thus, although the norm of organized skepticism is influenced by the

commercial reward system, I do not consider it here because of its limited empirical prevalence and thus its eclipsed empirical relevance to the study of commercialization in science. See Merton 1973.

3. Section C of the interview protocol in the Appendix.
4. Interview no. 44.
5. Interview no. 75.
6. Colyvas and Powell 2006; Stuart and Ding 2006; Etzkowitz 1989.
7. Interview no. 52.
8. Interview no. 87.
9. Shibutani 1962.
10. Interview no. 41.
11. Interview no. 72.
12. Interview no. 82.
13. Interview no. 91.
14. Interview no. 52.
15. Interview no. 24.
16. Merton 1973.
17. Cole and Cole 1973 and Gaston 1978.
18. Interview no. 36.
19. Interview no. 44.
20. Interview no. 72.
21. Interview no. 53.
22. See also Whittinger and Smith-Doerr 2005 and 2008.
23. Interview no. 29.
24. Interview no. 32.
25. Interview no. 80.
26. Interview no. 64.
27. Interview no. 70.
28. Interview no. 33.
29. Interviewer no. 32.
30. Cole and Cole 1973, 68.
31. Hagstrom 1965.
32. The connections that peer review has to misconduct and commercialization are understudied. For an overview of research on peer review, see Johnson and Hermanowicz 2017.
33. Interview no. 92.
34. Interview no. 24.
35. Braxton, Proper, and Bayer 2011.
36. Interview no. 34.
37. Interview no. 53.
38. Interview no. 70.
39. Interview no. 40.
40. Interview no. 62.
41. Blumenthal et al. 1996.

42. Interview no. 89.
43. Interview no. 86.
44. Interview no. 30.

Chapter 2 · The Reconstruction of Meaning and Status in Science

1. Abbott 1981.
2. Interview no. 69.
3. For an expanded discussion of intraprofessional status and the organizing knowledge of professions, see Abbott 1981 and Merton (1973), who makes a similar argument about science in particular.
4. Merton 1957.
5. Interview no. 82.
6. Weber (1921) 1968.
7. It is not the case that traditionalist groups are unusually small, as the average in this study is consistent with the findings of Louis et al., whose study of chemical engineering and life science research groups at the fifty universities producing the most PhDs showed that the average group has seven to eight graduate and postdoctoral researchers. Nor is it simply that specialties explain the difference. For example, of the three theoretical chemists in the sample, the two traditionalists have groups of two and ten, whereas the commercialist group comprises twenty researchers. See Louis et al. 2007.
8. "Engineering Research Centers," National Science Foundation: http://www.erc-assoc.org. Downloaded August 2011.
9. Interview no. 40.
10. Interview no. 26.
11. For a detailed explanation of market and hierarchical governance structures, see Powell 1990. For a description of the conditions that give rise to contract arrangements with expert labor, see Barley and Kunda 2004.
12. Interview no. 27.
13. Interview no. 89.
14. Interview no. 70.
15. Interview no. 46.
16. Manners 1975 and Louis et al. 2007.
17. Interview no. 74.
18. Zuckerman 1977.
19. Cole and Cole 1973; Merton 1973; Abbott 1981.
20. Abbott 1981.
21. Interview no. 48.
22. Interview no. 74.
23. Interview no. 32.
24. Interview no. 41.
25. Researchers have previously suggested that the meaning of clients in academic science is ambiguous. Goode argued that "client" is not in the lexicon of academics

because the cause of learning, not individuals, is served in academe. Similarly, Braxton states that "the knowledge base of an academic discipline is the client of scholarly role performance" because the goal of role performance is the creation of knowledge. See Braxton 1999, especially 142, and Goode 1957.

26. Interview no. 41.
27. Interview no. 55.
28. Interview no. 34.
29. Cole and Cole 1973; Merton 1973; Zuckerman 1977; Hermanowicz 1998.
30. Interview no. 46.
31. Interview no. 54.
32. Interview no. 39.
33. Cole and Cole 1973.
34. Zuckerman 1988, 526.
35. Merton 1973. The *Matthew Effect* is a term sociologists employ to discuss the accumulation of advantage. Drawn from the parable of talents in the Bible, in common parlance can be understood as "the rich get richer and the poor get poorer."
36. Interview no. 92.
37. Interview no. 67.
38. Interview no. 34.
39. Interview no. 37.
40. Interview no. 44.
41. Interview no. 46.
42. Interview no. 87.
43. Interview no. 25.
44. Interview no. 27.
45. Ashforth and Kreiner 1999 and Hughes 1951.
46. Interview no. 84.
47. Interview no. 92.
48. Interview no. 46.
49. Interview no. 54.
50. Interview no. 26.
51. Interview no. 28.
52. Interview no. 32.
53. Interview no. 27.
54. Interview no. 69.
55. Interview no. 64.
56. Interview no. 81.
57. Abbott 1981.

Chapter 3 · Embracing and Avoiding Commercial Trajectories

1. Stuart and Ding 2006.
2. The analysis in this chapter is derived from Section B of the interview protocol (appendix). I asked commercialists to discuss what factors had initially motivated

them to commercialize their research. Given the extent of their commercial practices, the interviews focused on the initial commercial act, which constituted a critical turning point in their career. To address the same issue among traditionalists, I asked them to discuss to what extent, if any, they had considered or pursued commercial behavior over the course of their career.

3. Lortie 1975.
4. Slaughter and Leslie 1997; Slaughter and Rhoades 2004; Etkzowitz 1998.
5. Interview no. 29.
6. Interview no. 28.
7. Interview no. 66.
8. Interview no. 68.
9. Interview no. 41.
10. Interview no. 49.
11. Interview no. 73.
12. Interview no. 26.
13. Merton 1973.
14. Whittington and Smith-Doerr 2005 and 2008.
15. Murray and Graham 2007.
16. Interview no. 66.
17. Zuckerman 1977 and Hermanowicz 1998.
18. Interview no. 30.
19. Interview no. 82.
20. Interview no. 83.
21. Support for these conclusions can also be observed in Szelényi 2013, which shows that graduate students in commercial contexts are highly aware of the market and of the symbolic and social meanings of money.
22. See also Wainwright et al. 2006; Gieryn 1999; Lamont and Molnar 2002.
23. Johnson and Ecklund 2015.
24. Interview no. 26.
25. Interview no. 49.
26. Interview no. 30.
27. See Sennett 2008.
28. Interview no. 81.
29. Interview no. 44.
30. For a detailed discussion of presence and material consciousness, see Sennett 2008, 130–35.
31. Interview no. 81.
32. This finding is consistent with existing research. Studying "maverick" scientists who conducted research on contraceptives during this period. Adele Clark notes that most scientists did not view working on technology as legitimate research, or even as science at all. See Clark 2000.
33. Interview no. 47.
34. Interview no. 54.
35. For example, in their study of structural characteristics that influence scien-

tists' transition to commercialization, Stuart and Ding (2006) argue that commercialist co-workers, commercialist coauthors, the eminence of commercialist co-workers, publication, and institutional prestige are key attributes of work contexts that encourage scientists to commercialize their work.

36. Interview no. 42.
37. Interview no. 80.
38. Interview no. 32.
39. Interview no. 29.
40. Interview no. 50.
41. Interview no. 43.
42. Interview no. 46.
43. Interview no. 91.
44. Interview no. 25.
45. Interview no. 87.

Chapter 4 · Identity Work in the Commercialized Academy

1. Interview no. 82.
2. Giddens 1991 and Alvesson and Willmott 2002.
3. Giddens 1991.
4. DiMaggio 1997.
5. Mills 1940; Sykes and Matza 1957; Scott and Lyman 1968; Thompson 1991; Thompson and Harred 1992; Ashforth and Kreiner 1999; Pogrebin, Poole, and Martinez 1992.
6. Ashforth and Kreiner 1999.
7. Interview no. 74.
8. Interview no. 34.
9. Festinger 1957.
10. Having observed the importance of reframing to scientists forming companies for the first time, one need not assume these are recollections of a neutralization process or an ex post facto excuse. Even in the event that neutralizations do occur as after-the-fact rationalizations, they may function as moral release mechanisms that facilitate future departures from convention. That is, in learning how to neutralize and cope with the stigma associated with a preferred course of action, one's identification with and commitment to a role could persist unconstrained by the pressures of social control that were previously operative. See Minor 1980 and Hirschi 1969.
11. Mills 1951.
12. Interview no. 86.
13. Interview no. 41.
14. One scientist whose first company was being incorporated in close proximity to the interview date was not included in this count. Academic conventions in constructing a CV suggest that omitting commercial involvement from one's CV is commonplace and not limited to scientists in this study.
15. Ashforth and Kreiner 1999.

16. Interview no. 93.

17. For discussion of the social norm of reciprocity, see Gouldner 1960.

18. Interview no. 36.

19. Gibbons and Gerrard 1991.

20. To be sure, the question could be interpreted in a positive light. If there were any assumptions underlying it, one could be a belief that commercialization potentially enhances collegiality. For instance, material benefits may be viewed as enhancing the resources and reputation of a department or contributing to societal welfare.

21. Interview no. 29.

22. Interview no. 82.

23. Interview no. 61.

24. Interview no. 34.

25. Interview no. 53.

26. Mullaney 2001.

27. Durkheim (1897) 1951. Whereas Durkheim alluded to the notion of a "never identity," Ebaugh 1988 made it explicit. Ebaugh's work shows how in the process of becoming an "ex," an individual deemphasizes a prior identity by avoiding actions associated with an "old self." Building on this insight, scholars now argue that just as behaviors that "individuals *used to, but no longer, do* are central in determining identity, so are acts in which one has *never* engaged." See Mullaney 2001.

28. Interview no. 46.

29. The utility of this analytic approach may be understood by considering reactions to killing. In different places and times, killings may be defined as criminal or noncriminal. The focus on the acceptability of commercialization, much like the focus on the criminalization of killing, is to ask how an act is defined in one way in some circumstances, but not others. See Cooney 2009.

30. Interview no. 43.

31. Interview no. 27.

32. Interview no. 31.

33. Interview no. 24.

34. March, Peter. "Broader Impacts Criterion": http://www.nsf.gov/pubs/2007/nsf07046/nsf07046.jsp. Downloaded August 2011.

35. Interview no. 27.

36. Merton's formulation of ritualism as a mode of adaptation invokes the basis for such identity work: "It is, in short, the mode of adaptation of individually seeking a *private* escape from the dangers and frustrations which seem to them inherent in the competition for major cultural goals by abandoning these goals and clinging all the more closely to the safe routines and institutional norms." See Merton 1957.

37. Interview no. 47.

38. Interview no. 68.

39. Interview no. 69.

40. Interview no. 46.

41. Interview no. 70.

Conclusion · Commercialism, Rationalization, and Fragmentation in Science

1. Cole and Cole 1973, 83–84.
2. Kalberg 1980.
3. Ibid.
4. Further evidence of the goal of controlling societal uncertainty is found in a report of the National Research Council called *The New Biology Initiative*. The NRC charged a panel of members from the National Academies of Science and Engineering and the Institute of Medicine to address the following question: "How can a fundamental understanding of living systems reduce uncertainty about the future of life on earth, improve human health and welfare, and lead to the wise stewardship of our planet?" See National Research Council 2009.
5. Freidson 1984.
6. Although not a major finding of the study, patterns in the data suggest that the use of technology in commercially oriented research may also contribute to efficiency. DNA-sequencing technologies provide an example from biological science, whereas combinatorial chemistry technologies represent similar tool use in chemistry. We may recall the commercialist who discussed searching through a data stream of "gibberish" for information with which he could make money, whereas others referenced "churning" and "canvassing" data without a hypothesis. Such processes are facilitated through technology that accelerates the rate by which data may be processed. Thus, efficiency is derived from the fact that aspects of the search for technologies may be automated.
7. One might argue that traditionalism includes notions of calculability because scientists count numbers of papers, or citations, but the distinction between the two rests in their *ends*. Traditionalists stress the *intrinsic* value of research and are thus not concerned with calculating the value of truth, such as the financial value of the discovery of a distant planet.
8. Tuchman's (2009) ethnography of a university stresses a "regime of accountability" that is operative through audit culture and surveillance of the professoriate, which encourages calculability through the quantification of scientific achievement. Discussing how university efforts to commercialize innovations "have undergone a progressive rationalization," Geiger and Sá (2009) point to predictability and calculability. Predictability emerges in various places throughout their analysis, including discussions of targeted investments, New York's Strategically Targeted Academic Research Centers, and corporate funding, "which has run strongly in the direction of greater focus and more explicit links with benefits for the corporation." Geiger and Sá refer to notions of calculability in their discussion of state officials, who "hope for tangible gratification from programs in the form of new business and job creation." Bok makes the case that universities have a lot to learn from business, including ways to achieve greater efficiency. Tuchman, 2009. For their discussion of targeted investments, see Geiger and Sa 2009, 114; discussion of the STAR Center is found on 93; see 58 for the discussion of corporate funding. Geiger and Sa 2009, 79; Bok 2003, 25.
9. Berman 2011.

10. Ritzer and Walczak 1988.

11. Levine 2005, 116.

12. Hermanowicz's books (2009) and (1998), which focus solely on the traditionalist reward system, demonstrate that differentiation of the academic profession occurs across three contexts of science: elite, pluralist, and communitarian. The role set of science is critical to this conceptualization because moral orders are organized around valued actions and beliefs: elites emphasize research in the presence of teaching; communitarians emphasize teaching in the presence of research; and pluralists are relatively committed to both aspects of the academic role. When we consider the influence of *competing* reward systems on the constitution of moral orders, we see that fissures of differentiation in science occur not only between roles but also within them. That is, two elite scientists within the same department may be characterized by drastically different roles and identities, such that the career paths they pursue diverge more than they overlap.

13. Clark 1983.

14. Baldwin and Chronister 2001.

15. Weber (1921) 1968; Ridgeway and Walker 1995; Goffman 1967.

16. Ridgeway 2006.

17. Durkheim (1951) 1897.

18. Hermanowicz 2011; Braxton 1993 ; Hackett 1994; Schuster and Finkelstein 2006, 87; Johnson, Vaidyanathan, and Ecklund, forthcoming.

19. If inhabitants of pervasive commercial contexts appear to acknowledge the presence of a status hierarchy, a question that follows is whether such status beliefs spread widely within the academic community. The disparity between funding for commercially targeted science and fundamental science may create a general recognition that commercialists are more valued. The concentration of commercialists in elite research universities may also perpetuate commercial attainment as a status belief. Belief formation in local contexts likely occurs through professional socialization and identity work. If doctoral students see that commercialists have larger lab operations and more funding for research, they may associate such dimensions of influence with their advisor's commercial status. Even graduate students trained under traditionalists could form similar beliefs. What matters is that individuals in the process of socialization are exposed to situations that lead them to develop the view that commercial status is associated with higher standing and power. Identity construction processes examined in chapter 3 similarly contribute to the diffusion of these beliefs. Social comparison, for example, seeks to project the superiority of commercialism over traditionalism by disparaging referents of traditionalist science. Interactively, therefore, identity construction processes attempt to persuade individuals of the favorable nature of commercialization in a status hierarchy.

20. Abbott 1988, 82.

21. Schuster 2011.

22. Slaughter and Rhoades 2004.

23. Godlee 2007. The other "milestones" included penicillin, x-rays, tissue culture, ether (anesthetic), public sanitation, germ theory, evidence-based medicine, vaccines,

computers, oral rehydration therapy, DNA structure, monoclonal antibody technology, smoking health risk.

24. Chemical & Engineering News. "List of Top Pharmaceuticals": http://www.pubs.acs.org/cen/coverstory/83/8325/8325list.html. Downloaded August 2011.

25. Boldrin and Levine 2008.

26. Litan and Cook-Degan 2011.

27. Kordal and Guice 2008.

REFERENCES

Abbott, Andrew D. 1981. "Status and Status Strain in the Professions." *American Journal of Sociology* 86 (4):819–35.

———. 1988. *The System of Professions: An Essay on the Division of Expert Labor*. Chicago: University of Chicago Press.

Aldrich, Howard, and Martin Ruef. 2006. *Organizations Evolving*. London: Sage Publications.

Alexander, F. King. 2001. "The Silent Crisis: The Relative Fiscal Capacity of Public Universities to Compete for Faculty." *Review of Higher Education* 24:113–29.

Alvesson, Mats, and Hugh C. Willmott. 2002. "Identity Regulation as Organizational Control: Producing the Appropriate Individual." *Journal of Management Studies* 39 (5):619–44.

American Association of Medical Colleges. 2002. "Protecting Subjects, Preserving Trust, Promoting Progress II: Policy Guidelines for the Oversight of an Institution's Financial Interest in Human Subjects Research." Retrieved April 2010 from http://www.aamc.org/members/coitf/2002coireport.pdf.

Ashforth, Blake E., and G. E. Kreiner. 1999. "'How can you do it?' Dirty Work and the Challenge of Constructing a Positive Identity." *Academy of Management Review* 24:413–34.

Association of American Universities. 2001. *Report on Individual and Institutional COI: Task Force on Research Accountability*. Washington, DC: Association of American Universities.

Azoulay, Pierre, Waverly Ding, and Toby Stuart. 2007. "The Determinants of Faculty Patenting Behavior: Demographics or Opportunities?" *Journal of Economic Behavior and Organization* 63:599–623.

Baldwin, Roger G., and Jay L. Chronister. 2001. *Teaching without Tenure: Policies and Practices for a New Era*. Baltimore: Johns Hopkins University Press.

Barley, Stephen R., and Gideon Kunda. 2004. *Gurus, Hired Guns, and Warm Bodies: Itinerant Experts in a Knowledge Economy*. Princeton: Princeton University Press.

Bekelman, Justin. E, Yan Li, and Cary. P. Gross. 2003. "Scope and Impact of Financial Conflicts of Interest in Biomedical Research: A Systematic Review." *Journal of the American Medical Association* 289:454–65.

Bercovitz, Janet, and Maryann P. Feldman. 2008. "Academic Entrepreneurs: Organizational Change at the Individual Level." *Organization Science* (19) 1:69–89.

Bercovitz, Janet E., Maryann P. Feldman, Irwin Feller, and Richard Burton. 2001. "Organizational Structure as Determinants of Academic Patent and Licensing Behavior: An Exploratory Study of Duke, Johns Hopkins, and Pennsylvania State Universities." *Journal of Technology Transfer* 26 (1–2):21–35.

Berman, Elizabeth Popp. 2011. *Creating the Market University: How Academic Science Became an Economic Engine*. Princeton: Princeton University Press.

Blumenthal, David, Nancyanne Causino, Eric Campbell, and Karen Seashore Louis, 1996. "Relationships between Academic Institutions and Industry: An Industry Survey." *New England Journal of Medicine* 334:368–73.

Bok, Derek. 2003. *Universities in the Marketplace: The Commercialization of Higher Education*. Princeton: Princeton University Press.

Boldrin, Michelle, and David Levine. 2008. *Against Intellectual Monopoly*. Cambridge: Cambridge University Press.

Bourdieu, Pierre. 1984. *Homo Academicus*. Stanford, CA: Stanford University Press.

Braxton, John. (1993). Deviancy from the Norms of Science: The Effects of Anomie and Alienation in the Academic Profession. *Research in Higher Education* 34 (2): 213–28.

Braxton, John. 1999. *Perspectives on Scholarly Misconduct in the Sciences*. Columbus: Ohio State University Press.

Braxton, John, Eve Proper, and Alan E. Bayer. 2011. *Professors Behaving Badly: Faculty Misconduct in Graduate Education*. Baltimore: Johns Hopkins University Press.

Cho, Mildred K., Ryo Shohara, Anna Schissel, and Drummond Rennie. 2000. "Policies on Faculty Conflicts of Interest at US Universities." *Journal of the American Medical Association* 284:2203–8.

Clark, Adele. 2000. "Maverick Reproductive Scientists and the Production of Contraceptives, 1915–2000+." In *Bodies of Technologies: Women's Involvement with Reproductive Medicine*, edited by Ann Rudinow Saetnan, Nelly Oudshoorn, and Marta Kirejczyk, 37–89. Columbus: Ohio State University Press.

Clark, Burton R. 1983. *The Higher Education System: Academic Organization in Cross-National Perspective*. Berkeley: University of California Press.

Cole, Jonathan R., and Stephen Cole. 1973. *Social Stratification in Science*. Chicago: University of Chicago Press.

Collins, Randall. 1975. *Conflict Sociology: Toward and Explanatory Social Science*. New York: Academic Press.

Colyvas, Jeannette A., and Walter W. Powell. 2006. "Roads to Institutionalization: The Remaking of Boundaries between Public and Private Science." In *Research in Organizational Behavior*. Vol. 27, edited by Barry Staw, 305–53. New York Elsevier.

Cooney, Mark. 2009. *Is Killing Wrong? A Study in Pure Sociology*. Charlottesville: University of Virginia Press.

Council on Governmental Regulations. 1999. "The Bayh-Dole Act: A Guide to the Law and Implementing Regulations." Washington, DC: Council on Government Regulations.

Di Gregorio, Dante, and Scott Shane. 2003. Why Do Some Universities Generate More Start-ups than Others?" *Research Policy* 32 (2):209–27.

DiMaggio, Paul. 1997. "Culture and Cognition." *Annual Review of Sociology* 23 (1): 263–87.

Durkheim, Emile. (1897) 1951. *Suicide*. Translated by John A. Spalding and George Simpson. New York: Free Press.

———. (1915) 1995. *The Elementary Forms of Religious Life*. Translated by Karen E. Fields. New York: Free Press.

Ebaugh, Helen. 1988. *Becoming an Ex: The Process of Role Exit*. Chicago: University of Chicago Press.

Etzkowitz, Henry. 1983. "Entrepreneurial Scientists and Entrepreneurial Universities in American Academic Science." *Minerva* 21 (2):198–233.

———. 1989. "Entrepreneurial Science in the Academy: A Case of the Transformation of Norms." *Social Problems* 36 (1):14–29.

———. 1998. "The Norms of Entrepreneurial Science: Cognitive Effects of the New University-Industry Linkages." *Research Policy* 27:823–33.

———. 2010. *The Triple Helix: University-Industry-Government Innovation in Action*. New York: Routledge.

Festinger, Leon. 1957. *A Theory of Cognitive Dissonance*. Evanston, IL: Row, Peterson.

Fleming, Lee, Charles King III, and Adam I. Juda. 2007. "Small Worlds and Regional Innovation." *Organization Science* 18:938–54.

Fleming, Lee, and Koen Frenken. 2007. "The Evolution of Inventor Networks in the Silicon Valley and Boston Regions." *Advances in Complex Systems* 10 (1):53–71.

Fligstein, Neil, and Doug McAdam. 2012. *A Theory of Fields*. New York: Oxford University Press.

Friedman, Joseph, and Jonathan Silberman. 2003. "University Technology Transfer: Do Incentives, Management, and Location Matter?" *Journal of Technology Transfer* 28:17–30.

Freidson, Eliot. 1970. *Professional Dominance*. Chicago: Aldine.

———. 1984. "The Changing Nature of Professional Control." *Annual Review of Sociology* 10:1–20.

Gaston, Jerry. 1978. *The Reward System in British and American Science*. New York: John Wiley & Sons.

Geiger, Roger L. 1988. "Milking the Sacred Cow: Research and the Quest for Useful Knowledge in the American University since 1920." *Science, Technology, and Human Values* 13 (3):332–48.

Geiger, Roger L., and Creso Sá. 2009. *Tapping the Riches of Science: Universities and the Promise of Economic Growth*. Cambridge, MA: Harvard University Press.

Gibbons, Frederick. X., and Meg Gerrard. 1991. "Downward Comparison and Coping with Threat." In *Social Comparison: Contemporary Theory and Research*, edited by Jerry Suls and Thomas A. Wills, 317–45. Hillsdale, NJ: Lawrence Erlbaum Associates.

Giddens, Anthony. 1991. *Modernity and Self-Identity*. Cambridge: Polity Press.

Gieryn, Thomas F. 1999. *Cultural Boundaries of Science: Credibility on the Line*. Chicago: University of Chicago Press.
Glenna, Leland L., Rick Welsh, David Ervin, William B. Lacy, and Dina Biscotti. 2011. "Commercial Science, Scientists' Values, and University Biotechnology Research." *Research Policy* 40:957–68.
Godlee, Fiona. 2007. "Milestones on the Long Road to Knowledge." A special supplement to the *British Medical Journal*, 13 January 2007 (334, no. 7584): http:// www.bmj.com/cgi/content/full/334/suppl_1/s2.
Goffman, Erving. 1967. *Interaction Ritual*. Garden City, NY: Doubleday Anchor.
Göktepe-Hulten, Devrim, and Prashanth Mahagaonkar. 2010. "Inventing and Patenting Activities of Scientists: In the Expectation of Money or Reputation?" *Journal of Technology Transfer* 35:401–23.
Goode, William J. 1957. "Community within a Community: The Professions." *American Sociological Review* 34 (3):194–200.
Gouldner, Alvin W. 1960. "The Norm of Reciprocity: A Preliminary Statement." *American Sociological Review* 25 (2):161–78.
Grassley, Charles. 2008. "Payments to Physicians." *Congressional Record—Senate*. June 4. S5029–S5030.
Hackett, Edward J. 1990. "Science as a Vocation in the 1990s: The Changing Organizational Culture of Academic Science." *Journal of Higher Education 61* (3):241–79.
———. 1994. "A Social Control Perspective on Scientific Misconduct." *Journal of Higher Education* 65 (3):242–60.
Hagstrom, Warren. 1965. *The Scientific Community*. New York: Basic Books.
Henderson, Rebecca, Adam B. Jaffe, and Manuel Trajtenberg. 1998. "Universities as a Source of Commercial Technology: A Detailed Analysis of University Patenting, 1965–1988." *Review of Economics and Statistics* 80:119–27.
Henkel, Mary. 2005. "Academic Identity and Autonomy in a Changing Policy." *Higher Education* 49:155–76.
Hermanowicz, Joseph C. 1998. *The Stars Are Not Enough: Scientists—Their Passions and Professions*. Chicago: University of Chicago Press.
———. 2009. *Lives in Science: How Institutions Affect Academic Careers*. Chicago: University of Chicago Press.
———. 2011. "Anomie in the American Academic Profession." In *The American Academic Profession: Transformation in Contemporary Higher Education*, edited by Joseph C. Hermanowicz, 216–40. Baltimore: Johns Hopkins University Press.
Hirschi, Travis. 1969. *Causes of Delinquency*. Berkeley: University of California Press.
Hodson, Randy. 1999. "Organizational Anomie and Worker Consent." *Work and Occupations* 26 (3):292–323.
Hughes, Everett C. 1951. "Work and the Self." In *Social Psychology at the Crossroads*, edited by John H. Rohrer and Muzafer Sherif, 313–23. New York: Harper & Bros.
Johnson, David R., and Elaine Howard Ecklund. 2015. "Ethical Ambiguity in Science." *Science and Engineering Ethics*. doi:10.1007/s11948-015-9682–9.
Johnson, David R., and Joseph C. Hermanowicz. 2017. "Peer Review: From 'Sacred Ideals' to 'Profane Realities.'" In *Higher Education: Handbook of Theory and Re-*

search. Vol. 32, edited by John C. Smart and Michael B. Paulsen, 485–527. Dordrecht, NL: Springer.

Johnson, David R., Brandon Vaidyanathan, and Elaine Howard Ecklund, forthcoming. "Structural Strain in Science: Organizational Context, Career Stage, Discipline, and Role Composition." *Sociological Inquiry*.

Kalberg, Stephen. 1980. "Max Weber's Types of Rationality: Cornerstones for the Analysis of Rationalization Processes in History." *American Journal of Sociology* 85 (5):1145–79.

Kirp, David. 2003. *Shakespeare, Einstein, and the Bottom Line: The Marketing of Higher Education*. Cambridge, MA: Harvard University Press.

Kogan, Maurice. 2000. "Higher Education Communities and Academic Identity." *Higher Education Quarterly* 54 (3):207–16.

Kordal, Richard, and Leslie Guice. 2008. "Assessing Technology Transfer Performance." *Research Management Review* 16 (1):1–13.

Krimsky, Sheldon. 2003. *Science in the Private Interest: Has the Lure of Profits Corrupted Biomedical Research?* Lanham, MD: Rowman & Littlefield.

Lam, Alice. 2010. "From 'Ivory Tower Traditionalists' to 'Entrepreneurial Scientists'? Academic Scientists in Fuzzy University-Industry Boundaries." *Social Studies of Science* 40 (2):307–40.

———. 2011. "What Motivates Academic Scientists to Engage in Research Commercialization: 'Gold,' 'ribbon' or 'puzzle'?" *Research Policy* 40:1354–68.

Lamont, Michele, and Virag Molnar. 2002. "The Study of Boundaries in the Social Sciences." *Annual Review of Sociology* 28 (1):167–95.

Leidner, Robin. 2006. "Identity and Work." In *Social Theory at Work*, edited by Marek Korczynski, Randy Hodson, and Paul Edwards, 424–63. Oxford: Oxford University Press.

Levine, Donald. 2005. "The Continuing Challenge of Weber's Theory of Rational Action." In *Max Weber's Economy and Society*, edited by Charles Camic, Philip S. Gorksi, and David M. Trubek, 101–26. Stanford, CA: Stanford University Press.

Litan, Robert E., and Robert Cook-Deagan. 2011. "Universities and Economic Growth: The Importance of Academic Entrepreneurship." In *Rules for Growth: Promoting Innovation and Growth through Legal Reform*, edited by Carl J. Schramm, 55–82. Kansas City, KA: Ewing Marion Kauffman Foundation.

Lortie, Dan. 1975. *Schoolteacher: A Sociological Study*. Chicago: University of Chicago Press.

Louis, Karen Seashore, David Blumenthal, Michael E. Gluck, and Michael A. Stoto. 1989. "Entrepreneurs in Academe: An Exploration of Behaviors among Life Scientists." *Administrative Science Quarterly* 34:110–31.

Louis, Karen Seashore, Janet M. Holdsworth, Melissa S. Anderson, and Eric G. Campbell. 2007. "Becoming a Scientist: The Effects of Work-Group Size and Organizational Climate." *Journal of Higher Education* 78 (3):311–36.

Manners, George E. 1975. "Another Look at Group Size, Group Problem Solving, and Member Consensus." *Academy of Management Journal* 18 (4):715–24.

Merton, Robert K. 1957. *Social Theory and Social Structure*. New York: Free Press.

———. 1973. "The Normative Structure of Science." in *The Sociology of Science: Theoretical and Empirical Investigations*. Chicago: University of Chicago Press.

Metlay, Grischa. 2006. "Reconsidering Renormalization: Stability and Change in 20th-Century Views on University Patents." *Social Studies of Science* 36 (4): 565–97.

Mills, C. Wright. 1940. "Situated Actions and Vocabularies of Motive." *American Sociological Review* 5:904–13.

———. 1951. *White Collar*. Oxford: Oxford University Press.

Minor, William. 1980. "The Neutralization of Criminal Offense." *Criminology* 18: 103–20.

Mullaney, Jamie. 2001. Like a Virgin: Temptation, Resistance, and the Construction of Identities Based on "Not Doings." *Qualitative Sociology* 24 (1):3–24.

Murray, Fiona, and Leigh Graham. 2007. "Buying and Selling Science: Gender Differences in the Market for Commercial Science." *Industrial and Corporate Change* 16 (4):657–89.

National Research Council. 2009. A New Biology for the 21st Century: Ensuring the United States Leads the Coming Biology Revolution. Washington, DC: National Academies Press.

National Science Board. 2004. *Science and Engineering Indicators 2004*. 2 vols. Arlington, VA: National Science Foundation.

National Science Foundation. 2014. National Center for Science and Engineering Statistics, Higher Education Research and Development Survey. Arlington, VA.

Owen-Smith, Jason, and Walter W. Powell. 2001. "Careers and Contradictions: Faculty Responses to the Transformation of Knowledge and Its Uses in the Life Sciences." In *Research in the Sociology of Work*, edited by Steven Vallas, 109–40. New York: JAI / Elsevier Press.

Phan, P., and Donald S. Siegel. 2006. "The Effectiveness of Technology Transfer: Lessons Learned, Managerial and Policy Implications, and the Road Forward." *Foundations and Trends in Entrepreneurship* 2:77–144.

Pogrebin, Mark R., Eric D. Poole, and Amos Martinez. 1992. "Accounts of Professional Misdeeds: The Sexual Exploitation of Clients by Psychotherapists." *Deviant Behavior* 13:229–52.

Powell, Walter. 1990. "Neither Market nor Hierarchy: Network Forms of Organization. *Research in Organizational Behavior* 12:295–336.

Pressman, Lori. 2002. "Association of University Technology Managers Licensing Survey FY 2001." Oakbrook Terrace, IL: Association of University Technology Managers.

Rapoport, Alan I. 2006. "Where Has the Money Gone? Declining Industrial Support of Academic R&D." Arlington, VA: National Science Foundation. NSF 06-38.

Rhoades, Gary. 2014. "Extending Academic Capitalism by Foregrounding Academic Labor." In *Academic Capitalism in the Age of Globalization*, edited by Brendan Cantwell and Ilkka Kauppinen, 113–36. Baltimore: Johns Hopkins University Press.

Ridgeway, Cecilia. 2006. "Status Construction Theory." In *Contemporary Social Psy-*

chological Theories, edited by Peter J. Burke, 301–23. Stanford, CA: Stanford University Press.

Ridgeway, Cecilia, and Henry A. Walker. 1995. "Status Structures." In *Sociological Perspectives on Social Psychology*, edited by Karen Cook, Gary Fine, and James House, 281–310. New York: Allyn & Bacon.

Ritzer, George, and David Walczak. 1988. "Rationalization and the Deprofessionalization of Physicians." *Social Forces* 67 (1):1–22.

Schuster, Jack. 2011. "The Professoriate's Perilous Path." In *The American Academic Profession*, edited by Joseph C. Hermanowicz, 1–17. Baltimore: Johns Hopkins University Press.

Schuster, Jack H., and Martin J. Finkelstein. (2006). *The American Faculty*. Baltimore: Johns Hopkins University Press.

Scott, Marvin B., and Stanford M. Lyman. 1968. "Accounts." *American Sociological Review* 33:46–61.

Sennett, Richard. 2008. *The Craftsman*. London: Penguin Books.

Shapira, Philip, Jan Youtie, and Sanjay Arora. 2012. "Early Patterns of Commercial Activity in Graphene." *Journal of Nanoparticle Research* 14:1–15.

Shibutani, Tamotsu. 1962. "Reference Groups and Social Control." In *Human Behavior and Social Processes: An Interactionist Approach,* edited by Arnold M. Rose, 7–27. Boston: Houghton Mifflin.

Siegel, Donald S., David Waldman, and Albert Link. 2003. "Assessing the Impact of Organizational Practices on the Productivity of University Technology Transfer Offices: An Exploratory Study. *Research Policy* 32 (1):27–48.

Siegel, Donald S., and P. Phan. 2005. "Analyzing the Effectiveness of University Technology Transfer: Implications for Entrepreneurship Education." In *Advances in the Study of Entrepreneurship, Innovation, and Economic Growth*, edited by Gary D. Libecap, 16:1–38. Bingley, UK: Emerald Group Publishing

Siegel, Donald S., R. Veugelers, and M. Wright. 2007. "Technology Transfer Offices and Commercialization of University Intellectual Property: Performance and Policy Implications." *Oxford Review of Economic Policy* 23 (4):640–60.

Skloot, Rebecca. 2010. *The Immortal Life of Henrietta Lacks*. New York: Crown.

Slaughter, Sheila. 2014. Foreword to *Academic Capitalism in the Age of Globalization*, edited by Brendan Cantwell and Ilkka Kauppinen, vii–x. Baltimore: Johns Hopkins University Press.

Slaughter, Sheila, and Larry Leslie. 1997. *Academic Capitalism: Politics, Policies, and the Entrepreneurial University*. Baltimore: Johns Hopkins University Press.

Slaughter, Sheila, and Gary Rhoades. 2004. *Academic Capitalism and the New Economy*. Baltimore: Johns Hopkins University Press.

Slaughter, Sheila, Scott L. Thomas, David R. Johnson, and Sondra Barringer. 2014. "Institutional Conflict of Interest: The Role of Interlocking Directorates in the Scientific Relationships between Universities and the Corporate Sector." *Journal of Higher Education* 85 (1):1–35.

Stephan, Paula. 2012. *How Economics Shapes Science*. Cambridge, MA: Harvard University Press.

Stephan, Paula E., Shiferaw Gurmu, A. J. Sumell, and Grant Black. 2007. "Who's Patenting in the University?" *Economics of Innovation and New Technology* 61:71–99.

Stuart, Toby, and Waverly Ding. 2006. "When Do Scientists Become Entrepreneurs? The Social Structural Antecedents of Commercial Activity in the Academic Life Sciences." *American Journal of Sociology* 112:97–144.

Sykes, Gresham, and David Matza. 1957. "Techniques of Neutralization: A Theory of Delinquincy." *American Sociological Review* 22:664–70.

Szelényi, Katalin. 2013. "The Meaning of Money in the Socialization of Science and Engineering Doctoral Students: Nurturing the Next Generation of Academic Capitalists?" *Journal of Higher Education* 84 (2):266–94.

Tajfel, Henri. 1982. "Social Psychology of Intergroup Relations." *Annual Review of Psychology* 33:1–39.

Tajfel, Henri, and John. C. Turner. 1979. "An Integrative Theory of Intergroup Conflict." In *Psychology of Intergroup Behavior*, edited by William G. Austin and Stephen Worchel, 33–47. Chicago: Nelson-Hall.

Thompson, John B. 1984. *Studies in the Theory of Ideology.* Berkeley: University of California Press.

Thompson, William. E. 1991. "Handling the Stigma of Handling the Dead: Morticians and Funeral Directors." *Deviant Behavior* 12:403–29.

Thompson, William E., and Jackie L. Harred 1992. "Topless Dancers: Managing Stigma in a Deviant Occupation." *Deviant Behavior* 13:291–311.

Thursby, Jerry.G., Richard Jensen, and Marie C. Thursby. 2001. "Objectives, Characteristics, and Outcomes of University Licensing: A Survey of Major U.S. Universities." *Journal of Technology Transfer* 26 (1–2):59–72.

Trice, Harrison. 1993. *Occupational Subcultures in the Workplace.* Ithaca, NY: ILR Press.

Tuchman, Gaye. 2009. *Wannabe U: Inside the Corporate University.* Chicago: University of Chicago Press.

United States Department of Health and Human Services. 2009. "How Grantees Manage Financial Conflicts of Interest in Research Funded by the National Institutes of Health." OEI-03-07-00700.

Vallas, Steven, and Daniel Kleinman. 2008. "Contradiction, Convergence and the Knowledge Economy: The Confluence of Academic and Commercial Biotechnology." *Socio-Economic Review* 6:283–311.

Wainwright, Steven P., Clare Williams, Mike Michael, Bobbie Farsides, and Alan Cribb. 2006. "Ethical Boundary-work in the Embryonic Stem Cell Laboratory." *Sociology of Health and Illness* 28 (6):732–48.

Washburn, Jennifer. 2005. *University, Inc.: The Corporate Corruption of Higher Education.* New York: Basic Books.

Weber, Max. (1921) 1968. *Economy and Society.* 3 vols. New York: Bedminster Press.

Whittington, Kjersten Bunker, and Laurel Smith-Doerr. 2005. "Gender and Commercial Science: Women's Patenting in the Life Sciences." *Journal of Technology Transfer* 30:355–70.

———. 2008. "Women Inventors in Context: Disparities in Patenting Across Academia and Industry." *Gender and Society* 22 (2):194–218.

Zucker, Lynne G., and Michael R. Darby, 2007. "Virtuous Circles in Science and Commerce," *Papers in Regional Science* 86 (3):445–70.

Zucker, Lynne G., Michael R. Darby, and Maryilynn Brewer. 1998. "Intellectual Human Capital and the Birth of U.S. Biotechnology Enterprises," *American Economic Review* 88 (1):290–306.

Zuckerman, Harriet. 1977. *Scientific Elite: Nobel Laureates in the United States*. New Brunswick, NJ: Transaction Publishers.

———. 1988. "The Sociology of Science." In *Handbook of Sociology*, edited by Neil J. Smelser, 511–74. Newbury Park, CA: Sage.

INDEX

Abbott, Andrew, 10–11, 47–48, 58, 79, 139
academic capitalism, theory of, 4–6, 11, 15
Academic Capitalism and the New Economy: Markets, State, and Higher Education (Slaughter and Rhoades), 4
Academic Capitalism: Politics, Policies, and the Entrepreneurial University (Slaughter and Leslie), 4
administrators, higher education, 5, 11, 33, 59, 144–145
advising, graduate, 41, 43, 83, 97–98, 107–108, 161n19. *See also* professional socialization
altruism, norm of, 93, 116, 134
American Association for the Advancement of Science, 18, 144
anomie, 138
Association of American Universities, 16–17
Association of University Technology Managers, 105
autonomy, professional, 4, 28; and commercialization policies, 139–141, 143, 146; and dirty work, 74; and identity work, 109, 114–115, 119, 123; and organizational reward systems, 33

Barringer, Sondra, 5
Bayer, Alan, 40
Bayh-Dole Act, 12, 17, 130, 140–142; and changes in professional socialization, 83–86, 89, 93, 95–96, 103
Becton Dickinson and Company, 61
Berman, Elizabeth Popp, 134
biotechnology, 82, 92, 140, 153n44
boards, corporate: efficiency and, 131–132; predictability and, 133; scientists' service on, 54, 61–62; university trustees and, 4. *See also* reference groups
Bok, Derek, 14, 160n8
boundary work: and division of labor, 52; and ethics, 82, 88, 91–93, 96
Bourdieu, Pierre, 14
Braxton, John, 40, 156n25
British Medical Journal, 141
British Petroleum Corporation, 61
bureaucratization, 50, 130

calculability: as a formally rational element of commercialism 4, 22; and material benefits, 88–89; and problem selection, 23–25; and tangibility, 94–96, 109–110, 129–130, 132–133, 134, 136–137, 160nn7–8; and visibility, 33, 46, 48, 50, 69, 79
California Institute of Technology, 96
California–San Francisco, University of, 97
California System, University of, 144
career paths, 11–12, 19–20, 81; attractors and, 88–96; facilitators and, 82–88, 96–103
Cech, Erin, 11
centers and institutes, research, 5, 28–29, 49–51, 55, 82–83, 132, 138, 160n8
Chemical and Engineering News, 141
Chronicle of Higher Education, 14
Clark, Adele, 158n32
Clark, Burton, 135
Cole, Jonathan, 37–38, 128
Cole, Stephen, 37–38, 128
commercialism, 3–4, 20, 129–135. *See also* calculability; control; efficiency; predictability
commercialist scientists, defined, 3, 6–7, 16–17

communalism: commercialization and, 21–23; defined, 7–9; espousal of, 38–45; and the institutional logic of science, 136–137
conferences, scientific, 39–40
conflict of commitment, 78, 99, 111, 138, 143, 145
conflict of interest, 38–39, 42–43, 78, 116, 143, 145
consulting, 31, 39, 61–62, 74, 91–92, 99, 102, 104
control: as a formally rational element of commercialism, 4, 20, 48; organization of work and, 50–53; reference groups and, 132, 134, 160n4; societal impact and, 62–64, 66, 79, 129–131, 136
curriculum vitae, 113, 158n14

Darwin, Charles, 63, 64
Defense, Department of, 77
development, economic, 2, 62, 64, 68–69, 131, 136, 140
DiMaggio, Paul, 104–105
Ding, Waverly, 157–158n35
dirty work, 73–79
disinterestedness: commercialization and, 21–23; defined, 7–9; espousal of, 23–30; as a facilitator of traditionalism, 96, 99; and the institutional logic of science, 136
dissonance, cognitive, 108, 110
Duke University, 1
DuPont Co., 61
Durkheim, Emile, 10, 118, 138, 159n27

Ebaugh, Helen Rose, 159n27
Ecklund, Elaine Howard, xi
Edison, Thomas, 63
Edwards, David, 1
efficiency: as a formally rational element of commercialism, 4, 22, 48, 129–132, 134, 136; laboratory organization and, 51–53, 79; peer review and, 39; reference groups and, 28–29; technology and, 160n6; universities and, 160n8
Einstein, Albert, 63
Eli Lilly and Company, 61
eponymy, 67, 136
equity interests, 90, 112
ethic, craft, 48, 93, 111
Etzkowitz, Henry, 8–9, 151n5

Feldman, Maryann, xii
fragmentation, 135–136
Freidson, Eliot, 14
funding, federal, 11, 34–35, 47, 55–56, 69, 71, 108–109, 139; of centers and institutes, 51; commercialization and, 24–25, 35, 39, 77, 93, 129, 130–131, 141, 142, 144. *See also* Defense, Department of; National Institutes of Health; National Science Foundation
funding, industrial: graduate students and, 41; identity and, 104–106, 118–119, 122; national level of, 2; professional control and, 132, 139–142; research bias and, 13, 27–28, 30

Geiger, Roger, 5
Genentech, Incorporated, 97
General Electric Company, 61
Georgia, University of, 1
Goffman, Erving, 21

Hare, Brian, 1
Hartwell, Lee, 25, 142
Harvard University, 1
Hermanowicz, Joseph, xi, 10, 161n12
higher education: corporatization of, 3, 140; social organization of, 4–6; social role of, 2–4, 11, 21, 24, 128
Hodson, Randy, 9
Hughes, Everett, 13
humility, 87

identity, 12–14, 15, 20, 104–106. *See also* identity work, techniques of
identity work, techniques of, 105; disavowal, 118, 119–121; disidentification, 20, 106, 117; professionalism, 115–117; reframing, 106–111; retreatism, 118, 125–126; ritual identification, 118, 121–125; role distancing, 111–113; social comparison, 113–115, 118–119
impact, societal: as an attractor to commercialism, 93; as a basis of status, 62–64, 134–135, 137; calculability and, 25, 94–95, 132–133; federal funding and, 121–125; as the goal of commercialism, 3, 22, 50, 76, 129–130; traditionalist pursuit of, 22, 25–26, 73; visibility and, 48, 68–72
institutionalism, theory of, 15
Intel Corporation, 61

jurisdiction, professional, 139–140

Kalberg, Stephen, 129–130
Kaswan, Renee, 1
Kirp, Daniel, 5
Kleinman, Daniel, 151n5
Krimsky, Sheldon, 7

Leidner, Robin, 13
Levine, Donald, 134
licensing agreements, 1, 4, 69, 88, 144; assigned to sampled universities, 16–17; assigned to study participants, 18–19
litigation, 1, 42, 97, 100
Lortie, Dan, 81
Louis, Karen, 56, 155n7

Matthew Effect, the, 67, 156n35
Merck & Co., 61
Merton, Robert K., 2, 7–8, 10–11, 14, 21, 23, 45, 48, 54, 87, 153n2
misconduct, 14, 92, 143, 154n32
Monsanto Company, 2, 61, 79
moral orders, conceptualization of, 10, 47–49. *See also* dirty work; status; visibility; work, organization of
moral taint, negation of. *See* identity work, techniques of

National Academy of Sciences, 18, 26, 67, 72, 126
National Institutes of Health, 34, 39, 77, 109, 115, 123–124
National Research Council, 160n4
National Science Foundation, xii, 51, 77, 122–124
Nature, 47, 68, 101
neutralization, techniques of, 106, 112, 158n10. *See also* identity work, techniques of
"never identity": defined, 117–118, 159n27; disavowal and, 121; retreatism and, 125
New Biology Initiative (National Research Council), 160n4
Newton, Isaac, 63
New York University, 144
Nobel Prize, 18, 57–58, 62, 64, 67, 83, 112
norms of science, 7–9; and career trajectories, 85, 87, 89, 99, 152n29; espousal of, 21–46; and identity work, 110–111, 114, 117; and intra-professional conflict, 135–136, 138. *See also* communalism; disinterestedness; organized skepticism; reciprocity, norm of; universalism

organized skepticism: commercialist reward system and, 153n2; defined, 7–9; and institutional logic of science, 136
Owen-Smith, Jason, 151n6

patents, 1, 12, 32, 42, 115, 137; assigned to sampled universities, 16–17; assigned to study participants, 18–19; efficacy of Bayh-Dole Act and, 141; novelty of, 121
peer review, 8, 40; market as a form of, 22, 25, 31–34, 46, 131; and secrecy, 23, 38–39, 154n32
Peifer, Jared, xi
Pfizer Inc., 61
politicians, 14, 55, 128, 134, 144
Powell, Walter, 151n6
predictability: as a formally rational element of commercialism, 4, 22, 129–130, 133–134, 136; industrial partnerships and, 28–29; organization of work and, 50; problem selection and, 23–24; targeted investments and, 160n8
priority in discovery, 2, 38, 40, 65, 91–92
problem selection, 22–26; calculability and, 130, 132; commercialization policy and, 142; societal impact and, 33
professional ideology: defined, 6; norms and, 8, 21, 45; status systems and, 46–47
professional socialization, 16, 50–51; commercialization policy and, 143–145; constraints to communalism and, 40–45; as a facilitator of commercialism, 82–84, 103; as a facilitator of traditionalism, 96–98, 103; identity work and, 107–111; societal impact and, 72. *See also* boundary work and division of labor; work, organization of
professions: conflict within, 14–15, 135–140; cultural dimensions of, 6–14
profit, 1–2, 19–20; commercial reward system and, 7–9, 75–77, 89, 92, 131; identity and, 116; in the medical profession, 134; and rarity of blockbusters, 1, 144; secrecy and, 43; traditionalist views of, 65–66. *See also* royalties
promotion and tenure, 8, 18–19, 30–31, 49; role of commercial activities, 31–38, 145
Proper, Eve, 40
publication, 17–18, 23, 38, 41–45, 47, 52, 121; of books, 75–76, 113–114; as a mundane accolade, 68–69, 94

purity, affirmation of. *See* identity work, techniques of
purity, professional, 47–48, 53–55, 58, 71, 79, 120. *See also* moral orders, conceptualization of
purity thesis, 10–11, 47–48, 58, 65, 79–80

rationality: bureaucratic, 129, 134–135; substantive, 20, 79, 129–130, 134
rationalization, 4, 130, 134–135; organization of work and, 50
rebellion, professional, 48, 53–55, 61, 63–65, 77, 79. *See also* moral orders, conceptualization of
reciprocity, norm of, 114
recognition: commercial, 2, 91, 93; scientific, 2, 23, 36, 38, 58, 62, 121, 136, 138
reference groups, 22, 26–30, 46, 85, 130–132, 142
research, interdisciplinary, 55
reward systems, conceptualization of, 2–4, 14–15, 161n12; organizational, 31–38, 153n54
Rhoades, Gary, 4–5, 140
Ritzer, George, 130, 134
role distancing, 105, 106, 111–113, 114, 117
royalties, 1, 32, 90, 93, 120, 134; distribution of, 31, 37, 88–89, 143–144; legitimacy of, 22, 34; status and, 138; visibility of, 60

Sá, Creso, 5
Science, 47, 52, 68
science: incremental, 75, 121; "job shop," 73–75
secrecy. *See* communalism
secrets, trade, 42, 45
Sennett, Richard, 93
Slaughter, Sheila, xi, 4–5, 7, 11, 140, 151n5
Smith-Doerr, Laurel, 87
Stanford University, 96, 117
startup companies, 1–2, 18–19; career burdens and, 97, 100; efficiency and, 29; reward systems and, 31; secrecy and, 42; status and, 61, 68
status, 57–58; client type and, 58, 61–62, 155n25; commercialist construction of, 62–64; hierarchies, 136–140, 161n19; income and, 58–61; traditionalist construction of, 65–66. *See also* purity thesis; visibility
stigma, commercialization and, 13–14, 76–79, 82, 85, 104. *See also* identity work, techniques of
stratification, 4, 5–6, 14, 67, 137–140
Stuart, Toby, 157–158n35

tangibility, 94–96; calculability and, 95, 130, 132–133
technology transfer offices: bureaucracy of commercialization and, 12, 88, 110, 133; commercialization policy and, 144–145; hiring of faculty and, 34; promotion of commercialization and, 5, 7, 11, 82, 105
Tesla, Nikola, 63–64
Thomas, Scott L., xii, 5
Thompson, John B., 6
traditionalism, ideology of, 6, 160n7; norms and, 7–8; status and, 10. *See also* norms of science; purity, professional
traditionalist scientists, defined, 3, 6–7
transformation of norms, hypothesis of, 8–9, 45–46
Tuchman, Gaye, 160n8

undergraduate education, 5, 9–10, 37, 77, 81, 109, 114, 143
United States Patent and Technology Office, 115
universalism: commercialization and, 21–23; defined, 7–9; espousal of, 30–38; and the institutional logic of science, 136
universities: Carnegie classification of, 16; linkages of, with the economy, 4–5, 82, 129; and management of commercialization, 143; public good and, 2, 8, 140, 144; scientists' perceptions of, 49, 57, 72, 116–117, 125, 138; as social arenas of struggle, 14–15; traditionalism in, 6

Vaidyanathan, Brandon, xi
Vallas, Steven, 151n5
Virginia, University of, 5
visibility: calculability and, 130, 133; commercialization as career burden and, 100–101; identity work and, 13; incremental science and, 75; moral orders of science and, 48; motivations of commercialism and, 12; reward systems and, 31–32; role distancing and, 111–113; status and, 58, 66–73, 136–137

Walczak, David, 134
Washington, University of, 144
wealth, displays of, 60–61, 77, 88, 112. *See also* role distancing
Weber, Max, 50, 129, 134

Welch, Jack, 63
Whittington, Kjersten Bunker, 87
work, cultural dimensions of. *See* career paths; identity; norms of science; status
work, organization of: commercialist, 49–53; traditionalist, 49–51, 53–57

Zuckerman, Harriet, 67